# Rebel Genius

Warren S. McCulloch's Transdisciplinary Life in Science

Tara H. Abraham

The MIT Press
Cambridge, Massachusetts
London, England

This book was set in Stone Sans and Stone Serif by Toppan Best-set Premedia Limited. Printed and bound in the United States of America.

Library of Congress Cataloging-in-Publication Data

Names: Abraham, Tara H.
Title: Rebel genius : Warren S. McCulloch's transdisciplinary life in science / Tara H. Abraham.
Description: Cambridge, MA : The MIT Press, [2016] | Includes bibliographical references and index.
Identifiers: LCCN 2016016507 | ISBN 9780262035095 (hardcover : alkaline paper)
Subjects: LCSH: McCulloch, Warren S. (Warren Sturgis), 1898-1969. | Neurophysiologists—United States—Biography. | Brain—Research—United States—History—20th century. | Cybernetics—United States—Biography. | Cybernetics—United States—History—20th century. | Neuropsychiatry—United States—Biography. | Mentoring in science—United States—History—20th century. | Engineers—United States—Biography.
Classification: LCC QP353.4.M33 A27 2016 | DDC 612.8092 [B]—dc23 LC record available at https://lccn.loc.gov/2016016507

10  9  8  7  6  5  4  3  2  1

For Joseph, Julian, and Martha
and in memory of Donna S. Abraham (1943–2013)

# Contents

# Acknowledgments

This book has been more than a decade in the making, and over the years I have accumulated intellectual and personal debts to many people. My work on McCulloch began as a single chapter in my doctoral dissertation, completed at the Institute for the History and Philosophy of Science and Technology (IHPST) at the University of Toronto. While that chapter bears little resemblance to this book, all of the scholarship I have produced since then has stemmed from it. While writing my dissertation, I benefited immensely from a very supportive community of scholars and students, and particularly from the mentorship and patience of my co-supervisors Paul Thompson and Mary P. Winsor.

In the early stages of this project I was fortunate enough to have two postdoctoral fellowships. My first postdoctoral stint, between 2000 and 2002, was at the Dibner Institute for the History of Science and Technology at MIT. Weekly colloquia and discussions at the Dibner were exciting and stimulating, and these meetings brought together sharp, constructive, and critical minds. I am grateful to the following people for their support during my time at MIT: David McGee, James Voelkel, Gordon McOuat, George Smith, Andrew Janiak, Abigail Lustig, Jutta Schickore, Jane Maienschein, Garland Allen, Anne-Marie Battis, Orna Harari, Annette Imhausen, Ben Weiss, and Carla Chrisfield. Between 2002 and 2004, I was a postdoctoral fellow in Department III at the Max Planck Institute for the History of Science in Berlin, led by Hans-Jörg Rheinberger. I thank Hans-Jörg for the opportunity to be a part of that incredible place and for his constructive criticism of my work. While in Berlin, my research also benefited from discussions and exchanges with many people, both inside and outside Department III, and I want to thank in particular Michael Hagner, Katharina

Schmidt-Brücken, Henning Schmidgen, Nancy Anderson, Leo Slater, Sofie Lachapelle, Antje Radeck, Uljana Feest, and Staffan Müller-Wille.

While teaching at York University between 2004 and 2006, the various threads of my research on McCulloch began to coalesce and as a scholar I was fortunate to be amidst an excellent group of colleagues, especially Ruthann Dyer, Kenton Kroker, Ernie Hamm, Suzanne Jaeger, Richard Jarrell, Katey Anderson, Joan Steigerwald, and Bernard Lightman. Katey, Joan, and Bernie in particular read an early iteration of my plan for this book and I am grateful for their constructive reading of my work during the very early stages of this project. I joined the Department of History at the University of Guelph in 2006, and I am indebted to all of my colleagues here for both their intellectual and moral support during the writing of this book, in particular, Christine Ekholst, Karen Racine, Bill Cormack, Elizabeth Ewan, Susannah Humble-Ferreira, Linda Mahood, Norman Smith, Susan Nance, Sofie Lachapelle, Catherine Carstairs, and Stuart McCook. I would also like to thank members of our department's summer work-in-progress group for their constructive feedback.

This book would not have been possible without a Standard Research Grant from the Social Sciences and Humanities Research Council of Canada, which supported my research between 2008 and 2011. Besides facilitating research trips to archives and conference travel, this support allowed me to enlist the research assistance of several talented graduate and undergraduate students: Calder Hutchinson, Christopher Laursen, Sarah Kemp, Jenna Healey, and Katherine Heyland.

In 2002–2003 I was awarded an American Philosophical Society (APS) Library Fellowship, which meant I got to spend the cold month of January 2003 in Philadelphia, Pennsylvania, poring over McCulloch's papers. I thank Corrina Burns for hosting me during that visit. I returned in May 2012 under much sunnier circumstances, and I am indebted to James Voelkel and Katy Fogle for putting me up during that trip. I thank the archivists I worked with at the APS Library for their assistance, particularly Valerie-Anne Lutz, Charles Greifenstein, and Earle Spamer. I also thank the archivists at the Massachusetts Institute of Technology's Archives and Special Collections, which I visited in 2002 and again in 2010, particularly Nora Murphy. Finally, I had splendid assistance from the archivists at the Rockefeller Archive Center in Tarrytown, New York, on two visits in 2005 and 2009, especially from Tom Rosenbaum. I am grateful to all for permission to quote from their documents and to use images from their collections.

In 2014–2015, during the final stages of writing this book, I had a sabbatical and spent a month at the Max Planck Institute for the History of Science under the support of Lorraine Daston's Department II. I also had a visiting position at the IHPST at the University of Toronto during that year. I thank both units for their support, particularly Muna Salloum and Denise Horsley at IHPST for all of their help.

Over the years I have benefited from the intellectual input, criticism, and encouragement of many people, and although I am likely forgetting someone, this list includes: Pnina Abir-Am, Ken Aizawa, Garland Allen, Robert Brain, Richard Burian, Carl Craver, Angela Creager, Uljana Feest, Peter Galison, Lisa Gannett, Slava Gerovitch, Yves Gingras, Christopher Green, Hunter Heyck, Evelyn Fox Keller, Manfred Laubichler, Willard McCarty, Everett Mendelsohn, Gualtiero Piccinini, Andrew Pickering, Gregory Radick, Ellen Rumm, Jutta Schickore, Maya Shmailov, and Rob Wilson. For more recent but very helpful discussions and exchanges on rhetoric, performativity, and identity, I thank Colleen Derkatch, Natasha Myers, and Jennifer Cypher.

For close reading of parts of the book in various iterations, I am indebted to Christopher Green, Kenton Kroker, Uljana Feest, Hunter Heyck, Mark Solovey, Susan Nance, Karen Racine, Susannah Humble-Ferreira, Danny O'Quinn, and Elizabeth Ewan. I am deeply grateful for the very constructive criticism I received from three anonymous reviewers, one of whom I now know to be Peter Cariani. While all of these readers gave me immensely helpful feedback, any errors, of course, are my own.

In the course of my work on this book had the very good fortune of being able to interview students and colleagues of McCulloch, whose openness captured the many ways he inspired them: the late Jerry Lettvin, Manuel Blum, and Lenore Blum. All three extended McCulloch's own lively spirit by very openly sharing their recollections and perspectives with warmth. McCulloch's family has also carried on his tradition of generosity, and this book would not have been possible without their faith in and support of the project. I spent a lovely weekend with McCulloch's eldest daughter, the late Taffy Holland, on the farm in Old Lyme, Connecticut, in 2002, where she and her brother David McCulloch warmly shared their views of their father's life and work. More recently, David has continued to communicate with me about his father, as has McCulloch's youngest daughter, Mary Jean Vasiloff. During the last stages of writing this book, McCulloch's granddaughter (Taffy's daughter) Anna Holland, along with David, helped find crucial photographs and ensured that I had access to high-quality images, which they graciously granted me permission to use. Their collective support of this work has meant the world to me.

At MIT Press, I have to first thank Margy Avery, who showed initial enthusiasm for the project in its very early stages, as well as Katie Persons. Although the reins were handed to Katie Helke later in the game, Katie has been a reliable editor and has ensured that what can be a long and convoluted process has gone smoothly. I also thank the editorial, production, and marketing teams at MIT Press, particularly my production editor Marcy Ross for her guidance and Bridget Leahy for her meticulous copyediting.

Portions of this book have appeared elsewhere in some form, particularly in the following publications: "'The Materials of Science, the Ideas of Science, and the Poetry of Science': Warren McCulloch and Jerry Lettvin," *Interdisciplinary Science Reviews* (2012) 37(3): 269–286; "Transcending Disciplines: Scientific Styles in Studies of the Brain in Mid-Twentieth Century America," *Studies in History and Philosophy of Biological and Biomedical Sciences* (2012) 43: 552–568; "Nicolas Rashevsky's Mathematical Biophysics," *Journal of the History of Biology* (2004) 37(2): 333–385; and "(Physio)logical Circuits: The Intellectual Origins of the McCulloch–Pitts Neural Networks," *Journal of the History of the Behavioral Sciences* (2002) 38(1): 3–25.

I have been blessed with many friends who, while they have always been huge supporters of my work as an academic, have also very helpfully reminded me of the important things that lie beyond academia. They are (in alphabetical order) Anne-Marie Battis, Janice Boich, Carole Burnett, Conor Burns, Helen Collins, Philippa Dowding, Thierry Dufaure, Jennifer Elliot Le Clainche, Marianne Fedunkiw, Catherine Gardner, Martina Gebhardt, Mary Giovinazzo, Rose Giovinazzo, Tina Giovinazzo, Felix Heeb, Leslie Jen, Paul Kemp, Mishann Lau, Jane Lawton, Jill Lazenby, Bonnie MacDonald, Heather MacDonald, Patrik Manzoni, David McGee, Ken Meiklejohn, Stefan Petri, Brigit Ramsingh, Jeelka Reinhardt, Megan Richards, Denyse Rodrigues, Tim Savage, the sopranos of The Toronto Choral Society, and Jeanette van Esbroeck.

My family has been a constant source of strength throughout this process, and I thank all of them from the bottom of my heart. My father, David

Abraham, was the first person in my life to encourage my intellectual aspirations, and while he at first raised his eyebrows at my choice of career, I know that he has come to appreciate the value of my chosen path. I also thank my sister, Sara Jane Ryan, her husband Brendan Ryan, and my nephew Jackson Ryan, for bringing so much laughter to my life. My parents-in-law, John F. Doane and Gerarda Kaye, and my brothers-in-law, Daniel Doane and James Dawson, have continually helped me keep things in perspective and lifted my spirits when they needed to be lifted.

I dedicate this book to my husband Joseph and my children, Julian and Martha, who carried on patiently and supportively while this book consumed much of my time and emotional energy. Joseph has tirelessly picked up the slack as a parent and has also been my biggest fan and sharpest critic. Finally, this book is also dedicated to the memory of my mother, Donna Shirley Abraham (née Shareski). Sadly she did not live to see it completed, but I know that she would be proud that it is finally finished.

# 1 Introduction

In 1960, at the age of sixty-two, Warren Sturgis McCulloch gave the Ninth Annual Alfred Korzybski Memorial Lecture in Fort Worth, Texas, where he told a story about a formative event in his intellectual development.[1] While a student at Haverford College in Pennsylvania in 1917, a teacher had asked him what he planned to do with his life. McCulloch said he hoped to answer the following question: "What is a number, that a man may know it, and a man, that he may know a number?" The first part of the question, McCulloch recalled, had been answered by the mathematicians. The second, more difficult part of the question—in essence, the problem of knowledge—was, according to McCulloch, the question that would direct his life's work.

McCulloch was not a philosopher in any institutional or professional sense. However, always an intellectual showman, he performed this identity carefully throughout the course of his scientific life. In the various disciplinary worlds through which he traveled—experimental neurology, psychiatry, neurophysiology, cybernetics, and engineering—McCulloch repeatedly (often retrospectively) cast his scientific work as addressing the problem of the mind and its relationship to the brain, and claimed that this common thread ran through the various disciplinary contexts and cultures in which he found himself.

That McCulloch told this apocryphal story at this particular juncture in his life is not surprising. By the early 1960s, the cybernetics movement was well underway and McCulloch had emerged as a public figure. He had chaired the ten Macy Conferences on Cybernetics that had taken place between 1946 and 1953, and had brought together a myriad of engineers, physicists, and biologists to explore the potential of using negative

**Figure 1.1**
Warren McCulloch in 1967. (Source: McCulloch Papers, Folder "McCulloch, Warren
S. 1950s–1960s Folder #2," Box 1, Series VIII Photographs, WSM Papers, APS.)

feedback as a way of making sense of natural and artificial systems. The ramifications of the cybernetics movement were already ubiquitous by the early 1960s—from the expansion of Cold War computing and information technologies to military-funded projects in man-computer symbiosis and problem solving.[2] In 1965, McCulloch had acquired a high enough profile that MIT Press published a collection of essays of his own choosing, entitled *Embodiments of Mind*. This was an exciting and transformative period in McCulloch's life, and he worked hard to construct both his own identity as a scientist-philosopher as well as an origin story for cybernetics—two practices that were closely intertwined.[3] Both projects involved a level of self-fashioning, a term coined by literary scholar Stephen Greenblatt in his work on Renaissance England and employed effectively by Mario Biagioli in his look at court culture in the time of Galileo.[4] Both of McCulloch's self-fashioning projects stressed unity and universality.[5] Just as McCulloch presented and defined cybernetics as a unified field that could transcend disciplinary boundaries, so he also presented himself as an intellectual who was driven by a philosophical problem and who rejected academic and disciplinary conventions—even his colleague and friend Ralph Gerard, in an obituary for McCulloch published in 1970, presented him as a "rebel genius."[6]

To be sure, autobiographical writings are of particular value to the historian who can only tell the story from the margins: autobiography tells the story from within.[7] Yet autobiographical writings are not transparent windows onto the past but ones that are colored with interpretations, strategies, and motivations. McCulloch's autobiographical recollections and origin stories for cybernetics cannot be taken as fact; they are reconstructions of his scientific life. McCulloch's career path was not especially radical, even if it followed a winding trajectory. He began his academic life as an undergraduate in philosophy and psychology at Yale in 1918, graduating in 1921. He received an M.A. in psychology from Columbia, and graduated from medical school there in 1927, with a focus in experimental neurology. After spending several years working at Bellevue Hospital in Manhattan, localizing areas of the brain involved in epileptic convulsions, McCulloch worked in the Admissions Service at Rockland State Hospital for the Insane in Orangeburg, New York, about twenty miles north of New York City. He returned to Yale in 1934 to work with primates in the laboratory of neurophysiology of Dutch-born neurophysiologist, Johannes Dusser de Barenne, on cerebral localization. In 1941, McCulloch went to the University of Illinois to join the newly formed Illinois Neuropsychiatric Institute in Chicago. Here he found his professional niche. At the same time, he began his work as a cybernetician, in particular his fruitful collaboration with Walter Pitts on the logic of neural networks. After over a decade at Illinois, as head of the neurophysiology laboratory, McCulloch went to MIT's multidisciplinary Research Laboratory of Electronics, where he spent the rest of his life addressing brain function in the framework of information processing.

By this time, studies of the brain in the American context had already begun to multiply and evolve from being restricted to highly empirical practices that stayed largely within the confines of traditional biomedicine to more multidisciplinary practices that involved theoretical modeling work in the tradition of physics, mathematics, engineering, and even philosophy. Regardless of how we interpret McCulloch's own self-fashioning, this shift is a central part of his legacy. From an early stage in his career, even before the cybernetics movement existed, McCulloch continually stressed the importance of theoretical modeling as the best way to understand an object as complex as the human brain. Today, we can witness the far-reaching effects of McCulloch's style of scientific practice in the contemporary fields of theoretical neuroscience and the cognitive sciences. These fields have traces of McCulloch's idiosyncratic way of approaching the brain and the mind: they transcend disciplinary boundaries (or erect new, more fluid boundaries), merge studies of the natural and the artificial, and are grounded in practices of theoretical modeling.

This book tells the story of McCulloch's scientific life. Rather than a strict biography, I frame the book as a variety of scientific biography that opens a window onto particular aspects of a past scientific age.[8] I use McCulloch's life and work to illuminate the complex transformations in scientific practice that took place in American brain and mind sciences during the twentieth century, in particular those surrounding the cybernetics movement. In order to connect McCulloch's scientific life to its wider contexts, I interpret his story through a focus on identity.[9] It is not worthwhile to blindly accept his identity as a cybernetician to the exclusion of all the other identities he adopted throughout his life. I do not deny that McCulloch was a maverick, and that he had wide-ranging, perhaps radical, ambitions. Yet as a scientific traveler, McCulloch performed various identities throughout his career that go beyond his professional roles: member of the post–World War I "lost generation," philosopher, poet, neurologist, neurophysiologist, neuropsychiatrist, collaborator, theorist, cybernetician, scientific administrator, modeler, mentor, and engineer. Ultimately, this book pulls apart the unified and universal rhetoric that McCulloch used in his self-fashioning and his construction of cybernetics to reveal a multiplicity of identities and a disunity to the cybernetic endeavor. Through his disciplinary trajectories and self-constructions, McCulloch embodies these features of cybernetics, for identity construction is a process that can play itself out at both the individual and disciplinary levels. Just as we can say the unity story for cybernetics is problematic, McCulloch's unity story is anything but unified.

Scientific biography, until quite recently, has received a bad rap among historians of science. In the wake of constructivist studies of science that emphasized the scientific process as contingent on social, cultural, and historical contexts, biographical treatments in the history of science were viewed as suffering from a hagiographical perspective that simply chronicled the lives of the "great men of science." More recently, however, many scientific biographies have rejected the empiricist framework and use their subject as a lens through which to examine the past critically.[10] Commentators point out the need to overcome the tension that pulls the historian between the biographical subject and his or her own agency, and the contingencies of historical context.

In order to dissolve this dualism, to balance historical and social context with a subject's own will, I draw on Judith Butler's framework of performative identity, in which gender is less a fixed attribute of individuals and more of a process, a doing, a becoming.[11] Here, identity is not fixed, rational, predictable, or stable, but dynamic and fluid.[12] Performativity helps negotiate the relations between McCulloch and his context via his activities as a scientist—which include practices both inside and outside the laboratory—experimental, theoretical, and rhetorical. Performativity is inclusive: as Karen Barad has argued, it overcomes the restrictions of many kinds of dualities: material and discursive, social and scientific, human and non-human, natural and cultural.[13] Just as a scientist does not hold up a mirror to nature, a biographer cannot ever reflect in a straightforward way a scientific life.[14]

McCulloch performed many of his identities through his scientific practices, that at once capture his own idiosyncratic way of doing science and the disciplinary worlds in which he operated. Yet McCulloch's life was more than a collection of performances. By the end of the 1960s, McCulloch's identities came together into the image of the seasoned philosopher, the polymath, the rebel genius. How did this happen? Was "the mind" really always at the core of McCulloch's scientific work?

I argue that McCulloch's scientific life at its heart was less a philosophical project and much more about transcending disciplines, the power of science to do away with metaphysics, and the power of a neurophysiological, biological psychiatry to eliminate dualist accounts of the mind and non-biological practices in psychiatry. Of course, one cannot be essentialist about McCulloch, despite his attempts to present his story as one of integrity.[15] Yet this faith in the power of science drove both McCulloch's work as a cybernetician and his work as a psychiatrist. McCulloch's medical and scientific concerns—understanding the biological basis of mental disorders—were of a piece with his more philosophical and theoretical cybernetic practices. Cybernetics ultimately became the framework through which McCulloch elaborated his ideas on the theoretical foundations of psychiatry. By examining the intermingling of McCulloch's disciplinary activities and the extent to which his clinical and experimental practices in neuropsychiatry and neurophysiology influenced his more theoretical work in cybernetics, this book offers a window onto the disciplinary cultures of both the brain sciences and cybernetics and the shifting dynamics of twentieth-century American science.[16]

## 1.1   McCulloch's Transdisciplinary Scientific Practices

One of the tools historians of science use to understand or make sense of scientific activity and identity is the academic discipline—its formation, emergence, consolidation, interaction, successes, failings, cultural power, and social power.[17] As an analytic framework, disciplines too can help the scientific biographer overcome the tension between individual agency and historical contingency; as Charles Rosenberg observed: "It is the discipline that ultimately shapes the scholar's vocational identity."[18] A scientific discipline, broadly understood, is the point of intersection of institutions, methods, object domains, and communities.[19] Disciplines are also the means through which consensus about scientific practice is reached and articulated, and in many ways McCulloch ultimately challenged the norms and standards that defined research in twentieth-century brain sciences. McCulloch's scientific life and the story of cybernetics serve as historical case studies of transdisciplinarity—an increasingly important feature of knowledge production since the mid-twentieth century.[20] McCulloch adopted multiple disciplinary identities throughout his career, both synchronously and chronologically. For most of his scientific life, McCulloch operated within traditional disciplinary frameworks, in prominent institutional milieux, and in many ways his scientific practices fit in nicely with disciplinary norms. As a result, his scientific practices aligned quite nicely with existing standards. Yet in other ways, particularly as a cybernetician, he attempted to transcend such cultures.[21]

I characterize McCulloch's scientific practices via three related features: *questions, modeling,* and *rhetorical strategies.* Each is useful in defining what was so unique about McCulloch and his brand of cybernetics, and each captures an important facet of his transdisciplinary strategies and those of the cyberneticians. McCulloch consistently frustrated and fascinated his contemporaries. Described as open and freewheeling, McCulloch had a broad scope and a tendency to ask big questions. In their goals of producing structural and functional knowledge about the cerebral cortex, traditional brain researchers pursued experimentally tractable research questions: for example, the precise relation of electrical recordings to the structural pattern of the cortex.[22] Much of this work was strongly mediated by the laboratory and clinical settings. Interests in precise mapping of brain functions shaped the questions they posed. McCulloch was certainly involved in such practices throughout his career, and indeed his work as a neurophysiologist during the 1930s and 1940s was grounded in experimental knowledge. However, as a cybernetician, he pursued knowledge of the brain by asking more ambitious questions that concerned theoretical models of perception, memory, and learning. He was deliberately provocative and irreverent, and he actively challenged accepted norms. Working with multiple identities, as a cybernetician he used the authority he gained in traditional disciplinary roles as a laboratory scientist and scientific administrator as a basis for posing big questions and making more ambitious claims about the brain and mind that bordered on the grandiose.

McCulloch asserted his success in addressing the problem of brain and mind via his practices of theoretical modeling, which were at the core of his scientific work as a cybernetician and beyond.[23] Particularly in his collaborations with Walter Pitts and Jerry Lettvin, McCulloch used abstractions and simplifications in his characterizations of brain function and organization. Members of traditional disciplines such as neurophysiology or neuropsychology, where knowledge of the brain's functional map was generated via precise instrumentation in a laboratory setting, did not always welcome these practices. The brain as a scientific object in these disciplines stood as part of a complex experimental system[24] that involved electrical instrumentation, probes, chemical solutions, surgical tools, laboratory animals, anesthesia, and all of the techniques and standards of experimental practice. Knowledge about the brain was strongly tied to such systems and to the localized, specific space of the laboratory.[25] In contrast, McCulloch's theoretical models only had a loose or at least a very different connection to such empirical knowledge. One of McCulloch's philosophical heroes was scientific philosopher Charles S. Peirce, who had described a third form of logical reasoning in addition to inductive and deductive: *abductive*, which involves, simply, "studying facts and devising a theory to explain them."[26] This was McCulloch's preferred mode of theorizing: while deductions lead from rules and cases to facts, and inductions lead to truth from cases and facts toward rules and generalizations, *abductions*, for McCulloch, lead from rules and facts to the hypothesis, which can then be tested by experiment.[27] McCulloch's "as-if" spirit was the antithesis of the precision that dominated brain research. I examine McCulloch's motivations for using theoretical models and the features of his modeling strategies that went against the grain of common practices in brain research by drawing on relevant philosophical literature that analyses modeling as a scientific practice.[28]

Finally, a prominent feature of McCulloch's scientific practices, and those of the cyberneticians, was his tendency to employ transdisciplinary rhetorical strategies and consequently a deliberately "loose" language—a technique that has been shown to accompany communication across disciplinary boundaries.[29] Cybernetic language, and McCulloch's in particular, was necessarily imprecise, a feature that frustrated its audience and critics. We may characterize the cybernetics group as operating in a variation of Peter Galison's notion of a "trading zone,"[30] yet problems of communication consistently plagued both McCulloch and the cybernetics group, in part because they were presenting what for many in biomedicine was a "foreign" language—that of mathematics. But other linguistic practices caused friction. Leah Ceccarelli has explored the rhetorical strategies scientists use to achieve interdisciplinary synthesis.[31] She identifies one such strategy—polysemy—as the practice of using textual constructions or concepts that have "many meanings."[32] As Ceccarelli notes, such strategies can be successful in bringing "different audiences, for different reasons, to accept a message." While such texts aim to be interpreted differently by different audiences, they are also meant to encourage collaboration and unity.[33] However, her work also suggests that controversies can surround interdisciplinary endeavors when the rhetoric employed by interdisciplinarians is expansionist in nature. The cybernetic vision ultimately was transformative and imperialist: problems in a variety of disciplines could be most fruitfully tackled using mathematical and modeling tools.

## 1.2   McCulloch and the Landscape of Twentieth-Century American Science

The challenge for the scientific biographer is to place McCulloch's scientific practices and the various identities he performed in his wider context. In keeping with performative identity, McCulloch was simultaneously a maverick and a product of his institutional milieux. While he challenged accepted norms, in many ways he occupied a legitimate and quite ordinary position in the landscape of American brain research. A close examination of his scientific life reveals that his institutional settings often fostered his fluidity and ease of movement between disciplinary cultures. With a career that spanned the period between the interwar years and the Cold War era, McCulloch was able to take advantage of new disciplinary configurations and patterns of patronage in American science in his pursuit of knowledge about the brain. Just as McCulloch operated within traditional disciplinary structures—even as these structures were evolving—and earned scientific credentials as a brain researcher, he challenged such structures. McCulloch's approach to the brain and mind can be seen as being shaped by three related facets of his scientific life and world.

First, McCulloch had a fascination from early on with mathematics and philosophy. This was in part shaped by the fluidity that existed between science and philosophy during the 1930s and the institutional context that McCulloch found himself in, particularly at Yale, both as an undergraduate and as a neurophysiologist. This affinity for—not expertise in—scientific philosophy was further developed during the 1940s as many members of the cybernetics movement mingled with members of the American unity of science movement. By the 1950s and 1960s, McCulloch was able to trade in on the scientific and administrative credentials he had earned during his career and make ambitious claims about the potential of cybernetics and theoretical modeling for solving the problem of brain and mind. The practice of theoretical modeling, often grounded in mathematics, became his scientific answer to a philosophical problem.

Second, McCulloch's distaste for psychoanalysis, a growing branch of American psychiatry especially during the 1940s and 1950s, drove him toward biological and scientific approaches as a neuropsychiatrist. Never prominent in clinical circles, McCulloch's brand of psychiatry was firmly rooted in the laboratory. He spent nearly two decades of his life studying the functional organization of the cerebral cortex of the brain, and this practice defined his identity as a neuropsychiatrist. In a world in which the psychiatric profession was being colonized by psychoanalysts, McCulloch stressed the importance of biological psychiatry, both in diagnosis and treatment. His rise to prominence as a neuropsychiatrist took place during the height of the cybernetics movement, and McCulloch's work as a cybernetician must be understood within the context of American psychiatry.

Third, McCulloch was able to tap into new patronage patterns in American science. While McCulloch's scientific life spanned the period between the interwar and Cold War eras, patrons in this period in various ways supported transdisciplinary and multidisciplinary practices. The Rockefeller Foundation, whose support McCulloch enjoyed during the 1930s and early 1940s, promoted a biological basis for psychiatry. During the period in which cyberneticians flourished, patterns of patronage in American science had already begun to shift dramatically in the wake of the Second World War. The postwar period emerged as one of close collaboration between the US government and natural and social scientists.[34] Physics and engineering have been the paradigmatic cases for examining this relationship; however a growing body of literature has addressed the implications of these developments for the biomedical sciences.[35] The rise of the

Office of Naval Research, the National Science Foundation, and the National Institutes of Health resulted a new federal structure for scientific funding, and McCulloch's work certainly benefited from this. While historians have focused on the consequences of this shift for the emerging fields of cognitive psychology and artificial intelligence, I address the influence of new funding patterns on studies of the mind's wetware: the brain.[36] Another important sponsor, the Macy Foundation, supported the interdisciplinary gatherings of the cybernetics group. And ultimately, following the Second World War, US government, particularly the military, financed much of the work that McCulloch's group did at MIT's interdisciplinary Research Laboratory of Electronics.

## 1.3   Historiography

Treatments of McCulloch's contributions to American science have often focused on his role as a cybernetician and his theoretical work on neural networks, and have emphasized his view of the brain as a digital computer, stressing the ways that this model has ushered in new understandings of what defines us as human beings and of our relationship to machines.[37] In the main, this historical work examines McCulloch and his legacy from a more contemporary vantage point that centralizes McCulloch's contributions to artificial intelligence, placing this identity front and center, interpreting everything else in his life—all that came before—as a steady path toward the inevitable. For historians, McCulloch was a key player at the Macy Conferences on Cybernetics (1946–1953) and in shaping cybernetic discourse.[38] Philosophers too emphasize McCulloch as an early advocate of a "computationalist" approach to the philosophy of mind.[39] Although Jean-Pierre Dupuy has emphasized McCulloch's role as a contributor to the conceptual foundations of artificial intelligence, he has not treated the relationship between this field and McCulloch's contributions in a historically rigorous way.[40]

In a perceptive analysis that rightly draws attention to McCulloch's work in the human and medical sciences, Lily Kay argued that McCulloch's work was central in the cognitivist turn that characterized the human sciences during the 1940s, when the mind became a legitimate object of scientific study.[41] Kay's argument fits in with a common narrative in the history of twentieth-century human sciences that states that prior to the so-called "cognitive revolution" that was underway by the 1960s, the mind as a scientific object was taboo in scientific psychology, and that following the revolution, the mind became a legitimate object of scientific study.[42] Processes such as learning, memory, and problem-solving were all were cast in terms of information processing and reference to internal psychological states.[43] Casting these changes in the language of "revolution" has justifiably been questioned recently, with recent work stressing the realignment of scientific practices and disciplines (for example, new cross-disciplinary alliances between psychology, linguistics, computer science, and neuroscience, rather than an overnight radical shift in understanding). However, McCulloch's work certainly forms a part of this story. My account of McCulloch, which aligns with Kay's, unpacks his stress on "mind" and the various meanings this term had in the diverse contexts within which he operated. I leave aside the more philosophical question of whether the mind is a scientific object in the universal sense, in keeping with Roger Smith's work on psychological categories as historical and social entities as opposed to natural kinds.[44] Rather, I take as a given the widespread view in history of science that scientific concepts are historically embedded in and are shaped by a complex network of epistemic goals (what knowledge are we after?) and investigative practices (what routes to that knowledge are privileged?) that define scientific communities.[45] "Mind" took on different meanings during the twentieth century for different groups of brain researchers, which often turned on cultural and scientific values. If we wish to place McCulloch in his context, we must acknowledge that his invocation of the term "mind" was idiosyncratic, and he was deliberate in the particular ways that he cast his scientific work in philosophical language.

While I acknowledge McCulloch's legacy for the cognitive sciences, I examine McCulloch through a different historical lens, one that brings into sharper focus the roles that McCulloch's particular scientific persona, his scientific practices, and his disciplinary and institutional milieux played in the broad scientific transformations of which he was a part. McCulloch in many ways embodies an important and as yet unexamined aspect of the cybernetics movement. Beyond an ontological blurring of the boundaries between organisms and machines, cybernetics was also about transcending disciplinary divides and promoting new ways of pursuing knowledge about the brain. A focus on McCulloch's life can tell us much about the disciplinary worlds he was a part of (brain sciences and cybernetics) and about standards and norms for making knowledge about the brain and mind.

Much scholarship in the history of cybernetics has focused on the far-reaching conceptual, cultural, and technological dimensions of the movement. Scholars have focused primarily on the political, military, and technological contexts of cybernetics,[46] its impact on the social sciences and economics,[47] the militarization of psychological knowledge during World War II and the Cold War through the development of control systems and weapons systems,[48] the cultural impact of cybernetics on our understandings of what it is to be human,[49] and the metaphorical and methodological ramifications of cybernetics for molecular biology and developmental biology.[50] Peter Asaro has focused on the electronic models of cyberneticians.[51] More recently, Andrew Pickering has drawn well-deserved attention to the British cyberneticians.[52] I build on Pickering's analysis in ways that highlight both the similarities and differences between the British and American contexts. Pickering's claim for British cybernetics—that its central object was the brain—is also true of the argument I present for McCulloch's brand of American cybernetics. However, my study differs from Pickering's in its focus on more epistemological questions of knowledge production rather than characterizing cybernetics as presenting a new ontology—in Pickering's view, not simply one which erases the distinction between organisms and machines, but one that is adaptive, performative, and nonpredictive. While Eden Medina's recent treatment of cybernetics similarly highlights the disunity of cybernetics, her analysis comes from a transnational perspective (i.e., between Chile and Britain).[53]

Overall, what has garnered less attention from historians of science are efforts by American cyberneticians such as McCulloch to transform scientific practice in an array of disciplines in the life and human sciences, and the complex ways these efforts were received and responded to by members of traditional disciplines. By understanding McCulloch's brand of cybernetics, I highlight features of the cybernetics movement that deserve more scholarly attention.

Because McCulloch's disciplinary homes were in neurophysiology and neuropsychiatry, and because he worked during a very transformative era for studies of the brain, this book also provides a much-needed contribution to historical literature on twentieth-century neurosciences. As Michael Hagner and Cornelius Borck noted in 2001 in the introduction to their valuable collection of essays on the neurosciences, the history of brain research in the twentieth century is still very much a "territory to be explored."[54] While several works have appeared since then that address the history of brain research from thematic, philosophical, and cultural perspectives,[55] the more general historical accounts of the history of twentieth-century brain research, while quite comprehensive, have been written from the perspective of participants in the field, and do not address the institutional or more social dimensions of scientific practice.[56] There do exist contextualist studies of brain research, however most emphasize the contexts of the nineteenth century.[57]

## 1.4   Overview

Chapters 2 and 3 not only chronicle McCulloch's scientific life from his college days to his first professional role as a neuropsychiatrist, they also provide an analysis of the disciplinary relations between science, medicine, and philosophy during the interwar period and the patrons that fostered such relations. Chapter 2 discusses McCulloch the student, particularly his days at Yale and at Columbia and the institutional conditions for the development of McCulloch's own perspective in psychiatry that simultaneously was grounded in experimental work on the brain as well as logic and philosophy. McCulloch aligned himself here with the "lost generation" and looked to science and technology to remedy a world shattered by war. Chapter 3 examines McCulloch's identity of neurophysiologist and his further work in cerebral localization at Yale during the 1930s, the Rockefeller Foundation's program for the scientization of psychiatry, and McCulloch's earliest exposure to scientific philosophy. The chapter also discusses the expansion of neurophysiology as a discipline and the experimental cultures that began to define knowledge-making about the brain during this period. While McCulloch's scientific identity here was firmly grounded in the laboratory, he continued to straddle the worlds of science and philosophy.

Chapter 4 explores the most transformative period in McCulloch's life: his early years at the Illinois Neuropsychiatric Institute at the University of Illinois at Chicago. This chapter turns a traditional episode in the history of cybernetics inside out, as it presents McCulloch's early collaborations with

Walter Pitts—often seen as a first iteration of "artificial intelligence"—as stemming from an emerging tradition in mathematical treatments of problems in the biomedical sciences as well as from McCulloch's theoretical position on clinical practice in neuropsychiatry. In addition to providing the institutional context for their collaboration and the role of the Rockefeller Foundation in fostering practices that bridged the gap between the mathematical and biomedical sciences, the chapter also highlights early reactions to McCulloch's disciplinary transgressions and his first public musings on how theoretical modeling could address the philosophical problem of mind and brain.

The next two chapters, when paired, examine McCulloch's professional identity as a neuropsychiatrist and his role as a cybernetician, arguing that these two identities are inseparable. Chapter 5 places McCulloch in the landscape of academic psychiatry during the 1940s and early 1950s, and explores the laboratory cultures of biological psychiatry during the World War II and postwar periods. Highlighting the role of patronage during this period, the chapter argues that psychiatry during this period was by no means unified, and that McCulloch occupied a minority position in the varied cultures of psychiatric research. His dissatisfaction and frustration at the growing influence of psychoanalysis drove his very biological framework for making sense of the diseased mind. Chapter 6 presents a picture of cybernetics as a disunified endeavor and highlights McCulloch's own "brand" of cybernetics as a way of addressing the problem of brain and mind. For all their talk of unity, cyberneticians were anything but unified. Ronald Kline has recently observed that in terms of scientific practice, cyberneticians offered diverse interpretations of what cybernetics was really about.[58] Further, if one takes even a cursory look at the transcripts from the Macy Conferences on Cybernetics, there is a marked lack of unity or consensus, as core members of the group attempted to present mathematical and theoretical models of brain function and the like to participants from medicine, anthropology, and the social sciences, and were met with expressions of confusion and frustration. This is the tension of cybernetics: cyberneticians often presented an anarchical, "anything goes" attitude, but at the same time also tried to present a unified vision. To highlight this tension, chapter 6 focuses on the reception of McCulloch's theory of the brain and mind on the part of traditional brain researchers and the disciplinary obstacles McCulloch faced.

Chapter 7 addresses simultaneously how, during the late 1950s and the 1960s, the brain acquired a new identity as a scientific object for McCulloch in the new institutional space of the Research Laboratory of Electronics at MIT, and McCulloch's new role as an engineer there. It highlights the brain as an information processor and how this new identity was facilitated by the complex multidisciplinary research practices fostered by patrons such as the Office of Naval Research. In this post-cybernetic context, the brain ceased to be straightforwardly an object of medical interest and became an object with multiple connections to communications, organization theory, and engineering. Along with this a new set of practices emerged for making sense of the brain that were simultaneously grounded in mathematics, philosophy, and theoretical modeling. As McCulloch's role as a psychiatrist was eclipsed by his identity as a engineer, the relevance and ramifications of his work on the brain expanded.

In chapter 8, I reflect on the scientific legacies of McCulloch's work in more recent practices that form part of the multidisciplinary field of cognitive science. McCulloch's own transdisciplinary practices, throughout his various disciplinary worlds, were products of both his contexts and his own agency. These practices, along with his own self-fashioning and the assessments of his peers and the public, helped to generate his final performative identity—the rebel genius.

# 2 The Student of Science, Medicine, and Philosophy

## 2.1 Introduction

Born at the turn of the twentieth century, Warren McCulloch came of age during a tumultuous period in American history. Drawn simultaneously to the arts and to the sciences from a young age, his early career path traversed the worlds of philosophy, psychology, experimental neurology, and psychiatry. By 1934, McCulloch appeared firmly grounded in the world of psychiatry, yet he was distinct in his philosophical perspective. In the face of an eclectic array of theories and practices, including Freudian psychoanalysis and the psychobiological theories of Adolf Meyer, McCulloch developed a framework where psychiatric conditions were understood in terms of logical relations. While this is arguably a remnant of his early interests in philosophy, McCulloch did not have any formal philosophical training beyond his undergraduate days. Yet his eventual perspective as a budding psychiatrist set him apart. What institutional and cultural conditions shaped McCulloch's unique perspective?

This chapter tells the story of how McCulloch became a transdisciplinary thinker. Much more than interdisciplinarity, which often involves working *at* disciplinary boundaries, transdisciplinarity involves *transgressing* such boundaries and "modes of disciplinary etiquette."[1] Recently Julie Thompson Klein has pointed to the varied meanings that transdisciplinarity can have according to context and motivations for being transdisciplinary—including integration, complexity, collaboration, system, and transformation.[2] Throughout his scientific life, McCulloch did not simply attempt to synthesize knowledge from different disciplines in an integrative way, but rather sought transcendence. By examining

McCulloch's formative academic years at Yale and Columbia against the backdrop of interwar science, medicine, and philosophy, I show that McCulloch capitalized on the fluidity that existed between these fields in American academic life. Disciplinary expansion defined both psychology and psychiatry during this era. Laboratory medicine was firmly established, and American philosophy was on the cusp of directly engaging with the sciences. The boundaries between these fields were easily traversed—between philosophy and psychology, between the laboratory and the clinic, between the sciences and the arts. As a young student with interests in all of these areas, McCulloch moved easily across the permeable borders that separated these fields and cultivated a transdisciplinary perspective. I discuss the institutional conditions for the development of McCulloch's views on the problem of the mind, which was at once grounded in experimental work on the brain as well as logic and philosophy.

McCulloch was also involved in a process of self-fashioning during this period, particularly through his poetry. It is problematic to read poetry as straightforwardly autobiographical, or simply as a window onto the past. Certainly, in a limited way, McCulloch's poetry can tell us not only about the context surrounding his coming of age, but also the ways in which he aligned himself with a particular generation moving through a period of scientific and cultural change. Unlike William Carlos Williams, another modern poet who happened to be a physician,[3] McCulloch wrote poetry that was peripheral to his scientific and medical work. However, his poetic work can be understood as a form of scientific practice and scientific self-expression. By performing the identity of a modernist poet, McCulloch deliberately and consciously placed himself as part of a generation. Rhetorically, it was important for him to connect with this particular group: he adopted the recognizable norms and values of a constructed "lost generation" reeling from the effects of World War I.[4]

McCulloch's poetic work opens up discrepancies among and between his scientific practices. His choice of poetry as a form of expression—particularly the sonnet—reveals much about his character and the inconsistencies and complexities of his various identities during this period. As a poetic form, the sonnet is rather traditional. Yet, at the same time, McCulloch was drawing on very modern themes—science, industry, and war. Ultimately McCulloch's identity and persona during the 1920s was comprised of paradoxes—irreverence and idealism—and wide interests—philosophical, medical, and scientific. Placing McCulloch in the fluid disciplinary landscape of post–World War I science and philosophy reveals that rather than being driven simply by the pursuit of the problem of knowledge, the constant in McCulloch's scientific life during this period, in all the various disciplinary worlds through which he traveled, was a faith in the power of science and philosophy to rationalize and remake a world broken by war and mental disease. While he drew on cultural markers of the postwar era—progressivism, scientism, the "lost generation"—and fashioned himself as part of this group, he set himself apart by following a radical path in psychiatry.

## 2.2  Early Childhood: Poet and Chemist

Warren Sturgis McCulloch was born on November 16, 1898, in Orange, New Jersey, an industrial town just northwest of Newark, to James William McCulloch (1857–1938) and Mary Hughes Bradley McCulloch (1865–1959). James had been born in Clifton, on Staten Island in New York City, and grew up there as part of a large family. The McCulloch family was of Scottish descent and had settled in Maryland before the American Revolution.[5] Mary had hailed from Maryland but was living in Washington, D.C., when she met James. The McCulloch family was upper middle class, as a 1910 census lists them has having a servant in the home. James McCulloch worked first as a bookkeeper and later as manager of the estate of Anson Phelps-Stokes, a New York City merchant and philanthropist.[6] The families were close, as Phelps-Stokes was invited to Warren's wedding to Ruth Metzger on May 29, 1924 (on which more below),[7] and James had been described as their "family man of affairs."[8] There were two other children in the family: a daughter Margaret Callender McCulloch (1901–1996), who eventually became a notable civil rights activist [figure 2.1], and Warren's older half-brother Paul, born to James's first wife in 1887. Warren and Margaret were very close growing up. They both attended the Beard School, a small private school about a half mile from their home.[9] Margaret recalled that they both learned quickly and did well at school, and both had a deep interest in poetry and literature. From an early age, Warren had an "eager, inquiring mind that went way beyond his assigned studies." He delighted in learning by heart poems such as Sir Walter Scott's *The Lady of the Lake*,

**Figure 2.1**
The McCulloch family in 1902. From left to right: Margaret, Mary, Warren, and
James McCulloch. Warren aged four. McCulloch Papers, Folder "McCulloch Family
1902–1903," Box 1, Series VIII Photographs, WSM Papers, APS.

Matthew Arnold's *Sohrab and Rustum*, and Edward Fitzgerald's translation of
*The Rubaiyat of Omar Khayyam*, a collection of four-line stanzas by the
twelfth-century Persian mathematician.[10] The two children loved playing
guessing games and charades, and stayed up past bedtime reciting poetry or
making up riddles. Summers were spent outdoors, either at rented homes in
Connecticut, on a 20-acre farm in Morristown, New Jersey, or on road trips
to the west coast.[11]

McCulloch attended high school at Carteret Academy, a private boys
school in Orange, where he excelled at Latin, did poorly at other languages
and spelling—reportedly being held back a year because of it—and simultaneously did well at mathematics and chemistry, setting up a basement
laboratory where, according to his sister, he conducted "endless experiments, some with happy results, others disastrous." Warren made friends
easily and was generous and warm-hearted, traits that some children took
advantage of. While Warren did not initiate fights, Margaret recalled, he
did have a fiery temper that would flare up when he was tormented or
mocked for his love of poetry. An idealistic child, Warren had aspirations to

the ministry, an interest likely shaped by the strong role that the Episcopalian church (an upper-middle-class branch of Protestantism) played in the McCulloch's family life: the family attended St. Andrew's Episcopal Church in South Orange. Grace at meals and nightly prayers were a regular part of life in the McCulloch household. According to McCulloch's son David, Mary, who was quite religious, pushed McCulloch to consider a career in the ministry.[12]

After graduating from Carteret Academy, McCulloch entered Haverford College in Haverford, a suburb of Philadelphia, in September 1917.[13] A small liberal arts college founded in 1833 on Quaker values, Haverford was originally established to attend to the educational needs of youth in the Society of Friends.[14] Education here reflected the rather liberal, progressive branch of Quakerism that was prominent in the eastern United States.[15] Sending McCulloch to Haverford likely meant little more than a quest for a "religious" education; however, the choice probably reflected a belief in the value of a broad education based on progressive values. For example, Quaker philosopher and writer Rufus M. Jones, who taught at Haverford, promoted a very liberal brand of Quakerism yet strove for unity among its followers.[16] McCulloch was enthralled with Jones, who encouraged McCulloch's interests in mathematics and philosophy. McCulloch apparently had an early interest in theology at Haverford—he later described himself during this period as a "prospective theologian"[17]—and took courses in physiology, English, French, Latin, mathematics, and history.[18] Outside of the classroom, McCulloch had diverse interests and hobbies. As his wife Rook recalled:

As a young man [McCulloch] found time to write poetry, to become a first-rate carpenter, to learn blacksmithing and welding. He had a very practical knowledge of surveying and a great deal of experience sailing and navigating. He read rapidly, remembered an enormous amount and, until the end of his days, studied and learned new things. Everything he learned was put to use and contributed to his remarkable sense of wholeness.[19]

Margaret's and Rook's recollections suggest a young man of wideranging interests and insatiable curiosity, even from an early age. McCulloch clearly saw the world from a broad perspective, and felt equally at ease with the more literary, cerebral world of poetry and literature as with the handson activities of carpentry and surveying.

McCulloch's arrival at Haverford took place only a few months after the United States entered World War I, following years of isolationism. Described by his sister as intensely patriotic and "no pacifist,"[20] McCulloch transferred to Yale University in September 1918, to join the United States Naval Reserve's Officer's Training Program for students, which was not offered at the Quaker Haverford College [figure 2.2].[21] Despite his entry into the Navy, McCulloch never saw active service (only 27 out of over 300 Yale students did). Yet his experiences at Yale were integral to McCulloch's self-presentation.

**Figure 2.2**
Warren McCulloch in the Navy at Yale University, 1918. McCulloch Papers, Folder "McCulloch, early years (ca. 1899–1918)," Box 6, Series VIII Photographs, WSM Papers, APS.

## 2.3    Undergraduate Days at Yale: Campus Life

With his coming of age during the 1920s, it would be tempting to place McCulloch within a generation of youth who had experienced the horrors of World War I—a "lost generation." Although recent historical work has problematized this neat characterization of the period, and has drawn attention to the complexities of the 1920s American experience according to race and class, it is safe to say that McCulloch was indeed part of the urban, white, middle-class, college-going, Protestant group who was swept up by the tumultuous social changes of the Jazz Age.[22] In 1918, McCulloch was Seaman (2nd Class) in the Naval Training Unit and the following fall was Boatswain's Mate (1st Class), spending several months at sea while continuing his courses.[23] Yale's Naval Training Unit had been established in 1916,[24] and in creating it, the US government likely had a similar motivation as Woodrow Wilson had in his creation of the Student Army Training Corps in July 1917: to combat low university enrolments due to the war.[25] The students in Yale's Naval Training Unit took courses in seamanship and navigation and devoted vacation periods to naval study, at times on board naval ships such as the *USS Massachusetts*.[26]

Despite having never experienced combat first-hand, McCulloch could not escape the indirect presence of the war at Yale. The war had transformed the campus and, by its end, Yale was essentially a military camp,[27] with the Naval Training Unit a prominent group on campus.[28] Many colleges throughout the United States participated in the war effort, connecting college life to the war effort seamlessly:

The graduate, dropping off at new Haven just now for a glance around the old University of his youth ... steps over to the College Campus only to find himself suddenly translated into a foreign country. He is directed out of Osborn Hall, now a Red Cross workers' quarters ... wanders into the basement of Wright Hall to be asked in one room whether he wants a physical examination by the local draft board lining up Class A 1 candidates for the Army ... sees in various directions batteries of artillery-men "hep-hepping" it in foot-drill or to Artillery Hall, and crosses Elm Street only to be stopped at the gates of Berkeley Oval by very spruce young men in blue uniforms and white hats who hold him up with a gun for his admittance pass.[29]

Given such an atmosphere, McCulloch could not have avoided the war's effects on American college life.

Paula Fass characterizes the 1920s as a period of adjustment, in which youth were both products of and agents of change.[30] The traditional historical picture presents youth of this period as cynical and disillusioned as a result of the Great War: they were, in this perspective, "iconoclastic, irreverent," and "frivolous" in their rejection of traditional beliefs and moral framework of the previous generation.[31] Yet World War I triggered both apathy and a small measure of enthusiasm—some of which bordered on patriotism.[32] McCulloch's college experience and processes of selffashioning illustrate elements of both characterizations.

At Yale, McCulloch majored in philosophy and gained a minor in psychology, graduating in June 1921. The yearbook for Yale's class of 1921 suggests that McCulloch was a popular, colorful character on campus. Descriptions of the activities and antics of the graduating class, by class historian Edwin Mims, Jr., include accounts of campus crazes such as top-spinning and roller skating:

... spring has brought an unexpected change—our minds have turned to thoughts of tops and roller-skates. There seems a new tie that makes all cliques kin. Through the length of Harkness [a large residential quad on campus], one thud follows another—a crash—and then, "Oh Hell! Why doesn't it spin!" It's the lamentation of those foolish virgins who failed to learn the gentle art of top-spinning in their youth.[33]

Mims went on to reflect, more seriously perhaps, that "[f]ew have stopped to think that in a month or so we can act this way only with the very real danger of being considered rather 'McCullochish'—if you know what I mean."

What might it have meant, for Mims, to be "McCullochish"? The comment followed the last section of the history of the graduating class, describing their hobbies in literary and colorful terms. Much of this captured the social atmosphere of the class—clubs, cliques, and sports teams. McCulloch, his son David recalled, could not carry a tune, but "loved to dance—even by himself ... he loved nonsense rhythms and plays on words."[34] McCulloch (whose nickname at Yale was "Drum," likely alluding to his participation in the university band) was probably infamous for a sense of fun and silliness—he received six votes from his graduating class for most popular campus character, and came in a close second in the vote for most original student.[35] He was likely not a star athlete, but a witty, popular student who, in an era of peer groups and conformity, was not afraid to be different.

Whatever social standing McCulloch had at Yale, like the entire class of 1921, his experience there was imbued with effects of the Great War. In the words of class secretary Willard D. Litt:

War and its aftermath have wrought an indelible impression on the life of our Class and have left a distinctive mark on our college years. Pleasant or otherwise as these memories may be, we can yet feel that in bowing to an inevitable duty we have done that, less than which could be expected of no Yale man. In years to come, it will be with a feeling of pride that we can look back on our short term of service.[36]

In this postwar climate, tops and roller skates would have been welcome distractions. As a product of wartime Yale, McCulloch seemed to inherit a strong sense of fun yet also duty and reverence. While McCulloch saw no active service, as an aspiring poet during this period, he explicitly aligned himself with the post–World War I generation.

## 2.4   The Poetry of War, Nature, and Science

McCulloch had been writing poetry since his early days as an undergraduate student at Haverford College, and continued to do so at Yale.[37] Although not straightforwardly autobiographical, there are limited ways in which we can glean some sense of McCulloch's experiences during this period and avoid the controversies surrounding autobiographical poetry.[38] In doing so it is crucial that we take these works not as truthful accounts of McCulloch's context but as performative acts, both in terms of the content of his poetry and the form that it takes. McCulloch's poems from this era do not simply attempt to chronicle his experiences nor do they confess personal secrets. Nor should they be viewed has simple containers of historical information about McCulloch and his context. Rather, they should be read as deliberate and reflexive constructions of an identity. In terms of content, the poetry is an exercise in aligning oneself with an era, a collective experience—in this case, modernity as it unfolded in the period after World War I. In the introduction to the collection of his poems, published in 1959, McCulloch states that they were the creation of the second "character" that defined McCulloch's life, who, in McCulloch's words, was "born of a war to make the world safe for democracy." This, of course is an allusion to President Woodrow Wilson's request for a Declaration of War against Germany in his April 2, 1917, speech to Congress.[39]

Yet McCulloch's choice of the sonnet reveals just as much about his persona during this period. As a poetic form, the sonnet originated in Italy with Petrarch during the middle ages, and flourished in Elizabethan England, of course, with Shakespeare. By the twentieth century, the sonnet was also used by modernist poets such as Robert Frost and W. H. Auden, while other modernists "escaped" the form in favor of "free verse." A sonnet is bounded by rules—traditionally, 14 lines, iambic pentameter, and a patterned rhyme scheme. It has been said to represent an aesthetic ideal.[40] Thus, while McCulloch's poetry of this period was quite modern in subject, it was classical in form.

There are really two things going on in McCulloch's poetry from this period. First, he highlights the sense of loss that plagued many during this period, and second, he connects himself to postwar progressivist thought in his reverence for science and technology. In dating this character as emerging from the Great War, McCulloch reminded his reader "of the world into which he was born aged twenty-one. The place is well this side of Paradise."[41] This, of course, is an allusion to F. Scott Fitzgerald's first novel, *This Side of Paradise* (1920), which examines the lives of American college youth following the war and shocked the contemporary public for its portrayal of their raucous behavior. The war brought a level of disillusionment to American youth, creating a cynical rejection of older moral values and conventions that had defined American culture before the war.[42] However, Fitzgerald also implicitly presents a yearning for the past, for former traditions, for stability.[43] McCulloch's reference to Fitzgerald helps him capture an important facet of life during the early 1920s for men of McCulloch's generation. If Fitzgerald portrays the youth culture of this period as full of gaiety, raucousness, and naïve optimism, it was always with the backdrop of war. As Paula Fass has observed, "Fitzgerald's heroes were gay but never light-hearted … it was gaiety based on a sense of loss. …"[44] Thus, McCulloch fashions himself as experiencing the loss of a golden past. The effects of war came with a perceived lost world; and McCulloch certainly identified with this:

In sad humility we learn of death,
Of lonely tears and half the world at war,
With our dimmed eyes we come again to read
The lesson that we should have learned of home,
Among familiar needs whose every breath,

Inspired with love's beatitude, far more
Than all the prophecies of all the years,
Was panting for the union of the race.
But we to little State and transient God
Gave all our souls and let our loved ones bleed.
Thus have we bought again the vanished grace
Of nature's moral law. Again we come
Out of our lesser loyalties, in tears,
To build love's well-earned city in the rich sod.[45]

From his post–World War II vantage point, McCulloch introduced this poem and the sequence of sonnets of which it was a part as "too easily dated after the war, while there was still hope (in the hopeful generation) for a league of nations. ... It is consoling to recall that we once felt so well about our one world. ..."[46] At the same time, there were also remnants of progressivist thought following World War I, and one way this manifested itself was in efforts toward reconstruction in the face of change.[47] For progressives such as George A. Coe and Ben Lindsey, American youth of this period were witnessing an industrial transformation.[48] Youth of the 1920s were coming of age in an era of "speed and science."[49] Although the war had shown that "machinery and science" without moral health are a "curse," science and technology when shaped by a moral framework are signs of progress.[50] Another of McCulloch's sonnets from this period more directly praises soldiers, but not before a tribute to scientists and engineers of the modern age:

Sing then to the men that make the road
And tunnels through the hills and bridges slung
Over the wide valley and the deep ravine,
Till streets are crowded with the whirling cars,
And the long-leveled track rumbles to the train.
Sing to the men that left some safe abode
And gazing fixedly on distant stars
Conquered the sea's tumultuous domain.
With oar and sail and steam, gallant they clung
To their wide quest, over unfathomed green
Bearing the timorous body and the heart's load,
Till now through space the wireless word is flung,
And the skilled pilot guides the soaring plane.
With these we win the world. To these my ode.[51]

War was chaotic, devastating, and destructive. Science had the power to bring order to the natural world thrown into chaos by global conflict. Many of McCulloch's reflections on the Great War are filled with sorrow for its horrific effects and reverence for those who fought in it. But many of these reflections are colored with the themes of the natural world, our place in it, and the roles of science and technology in remaking the world. Other modernist poets, besides experimenting with new forms, also similarly engaged with the modern world and the "machine-age metropolis."[52]

Besides reflecting McCulloch's sentiments on war and lost innocence, the poems capture McCulloch's intellectual aspirations and his faith in the power of science. During the 1920s, McCulloch clearly had a sense of optimism about the roles of science and philosophy to humanity in the postwar context:

Not with our loneliness, not with our dream,
Can our earth be taken and our heaven won,
But with our toil and ponderous reasoning,
And with our little changing of the cosmic stream.
Slow science patiently from law and fact
Designs the implements for laboring,
And love's proud home, our earth, which man
shall build,
A temple to the happiness he so long lacked.
Land and sea and sky, with careful thought
And toiling hand controlled, shall soon be filled
With the calm pulse of our united peace;
And Time and Space, no longer feared and fought
Drop back neglected in the course we run
To be forgotten ere our daylight cease.[53]

McCulloch was aligning himself with progressivism—not only the Wilsonian brand in which the American involvement in World War I contributed to the spread of democracy—but also in the sense of celebrating the potential of science, technology, and industry to rebuild. Unlike others of his generation, McCulloch did not use the experience of war as fuel for artistic expression or a rite of passage, but in a more directly progressivist spirit—as a rationale for the praise of modern industrial culture.[54] Although progressivist thought was on the wane following the war in America, science had acquired enormous prestige.[55] New machines and devices were everywhere, Einstein was front-page news, Frederick Winslow

Taylor's "scientific management" defined factory production, and psychology and psychiatry—with Freud a household name—reigned supreme in the decades following the war and were seen by many as the solution to society's problems. Science, in this context, was not simply a tool for the growth of knowledge or technological advancement, but an instrument of cultural and social change—in essence, science was at the heart of the modern world.[56] While the effects of war had an impact on McCulloch during this period, and there was, likely a level of irreverence in his character, this was tempered by a reverence for the power of science and reason to make sense of a world that no longer made any sense. Thus, McCulloch at once shows reverence for a traditional form yet takes liberties with its rules. The sonnet, although a strict, constrained, box-like form—like a puzzle one had to solve—was capable of being tweaked. While the above examples of McCulloch's sonnets stay within the confines of the 14-line structure, not every line was in the form of an iambic pentameter, nor did the structure take the form of a dual proposition and resolution, as many traditional sonnets did.

McCulloch's disillusionment and sense of loss brought about a crisis of faith—his faith in human nature was tested. In this context, he was drawn to rational, scientific approaches to the human mind. This may have simply been an attempt to elevate his poetry in a time when science and technology had cultural power, yet it is likely that it also reflected his scientific leanings.

## 2.5 Academic Experiences: Scientific Philosophy and Psychology

McCulloch majored in philosophy at Yale and gained a minor in psychology. Philosophy at the time was emblematic of the secularization and expansion that had defined the American educational system since the Civil War, as it tried to find its intellectual and institutional place in a post-Darwinian world of thought.[57] Most of McCulloch's coursework at Yale was in philosophy, with some English, German, history, and mathematics. His senior year had general chemistry with radiochemist Bertram Boltwood and biology with Lorande L. Woodruff; both courses had minimal laboratory work.[58]

Early twentieth-century philosophy at Yale was in the shadow of philosophy at Harvard and Columbia. While Yale philosophers such as

Charles Bakewell and William Ernest Hocking emphasized the civic duty of philosophers as contributors to "public culture,"[59] this was also an era when epistemology—the study of knowledge and how we acquire it—was at the core of philosophical study.[60] McCulloch's philosophical training took place in an era of high prestige for the sciences, and scientific philosophers saw scientific achievements having implications for the study of knowledge.[61] The two most dominant trends in scientific philosophy during this period were logic, in the tradition of Bertrand Russell and Alfred North Whitehead, and the pragmatism of Charles Peirce, William James, and John Dewey. McCulloch reported that while at Yale he had begun to read Peirce, Whitehead, and Russell.[62] In their *Principia Mathematica* (1910–1913), Russell and Whitehead had developed a system of logical notation for reducing mathematics to logic,[63] and Russell himself applied logic to the empirical world and believed epistemology could be reduced to purely logical relations.[64] Pragmatism was concerned with the primacy of the scientific method and also with the role of logic in scientific reasoning.[65] Historians of philosophy have argued that there were deep connections between scientific philosophy and contemporary science during the early part of the twentieth century.[66] On this reading, science offered resources for the practice of philosophy itself and the exploration of questions related to epistemological questions about objectivity. Dewey, for example, emphasized connections between philosophy, biology, and psychology.[67]

 History of philosophy was also on the menu in university training in philosophy, especially Immanuel Kant's *Critique of Pure Reason*.[68] Within this idealist framework, the mind and its operations occupied a central place. Scientific philosophers concerned themselves with the nature of human perception and aimed to connect logic to Kant's notion of the synthetic a priori. Pragmatists such as Peirce and James, who had both been trained in the sciences, adapted Kant in their theories of knowledge, which stressed, among other things, practical perspectives and especially the work of logicians and scientists.[69] Peirce, who became one of McCulloch's heroes, saw science and philosophy as mutually independent.[70] While James embraced science less wholeheartedly than Peirce, he was a "serial transgressor" of disciplinary boundaries, saw continuities between disciplines, and aimed to unify and simultaneously reconfigure knowledge and American society by expanding philosophy's scope.[71] Thus, despite institutional pulls that began to carve out a separation between science, psychology, and philosophy, McCulloch's undergraduate academic experiences were shaped by the kind of fluidity promoted by scientific philosophers.

For example, many American philosophers became concerned specifically with psychology as a science—that understanding mental life may gain us insight into the problem of knowledge and how we acquire it. These issues were prominent in William James's *Principles of Psychology* (1890), which presented psychology as a natural science (although in a much different vein than experimentalists). Psychology's sphere of inquiry, for James, should be to connect empirically the processes of thought with the attributes of the brain.[72] Although historically tied to philosophy, psychology was undergoing its own process of professionalization and scientization during this period, and by the mid-1920s existed as an independent discipline.[73] However, at Yale at least, psychology and philosophy, institutionally speaking, were not formally separated until 1928.[74] Thus, although his undergraduate thesis in psychology addressed a problem related to aesthetics, and used the methods of applied psychology by tackling the problem of perception, McCulloch operated within a rather fluid disciplinary landscape where epistemological questions about the nature of perception sat at the intersection of scientific and philosophical modes of inquiry.

According to the recollections of his sister, Margaret, McCulloch was drawn to psychology as a way of making sense of his mother's "condition": as a way of "interpreting to himself and to me the nature of my mother's problems and some of my own which stemmed from them ... my mother's condition may have been one of the factors contributing to this interest."[75] Two of McCulloch's children, David and Mary Jean, have no recollection of such a condition, and speculate that Mary may have had a form of dementia in her later years. Although it is impossible to determine if this was McCulloch's true motivation, in his final undergraduate year at Yale, McCulloch attended a series of lectures at Yale's art school given by the Canadian-born artist and teacher Jay Hambidge.[76] Hambidge's ideas would prompt McCulloch's undergraduate thesis on the psychology of human perception.

Trained as a professional illustrator, Hambidge was known for his work on a system of proportion called "dynamic symmetry" and its role in ancient Egyptian and Greek art and architecture. In his 1920 book, *Dynamic*

*Symmetry: The Greek Vase*, Hambidge argued that dynamic symmetry, characteristic of organic life and observable in organic forms like the shell, was the dominant principle of proportion and design. Dynamic symmetry was based on what are called root rectangles—those in which the relation between the *length* of the long side to the length of the short side is "incommensurable"—specifically, a ratio of $\sqrt{n}$ to 1. Hambidge theorized that ancient Egyptians and Greeks had developed a system of proportion and design through the manipulation of areas within such rectangles and the diagonal lines that cross them, rather than standard measurements of length. Most important for dynamic symmetry was the so-called "Whirling Square Rectangle" or "Golden Section," based on a ratio of $\sqrt{5}$ to 1, which Hambidge purported contained a perfect sense of proportion and thus was most aesthetically pleasing to the viewer. Because images based on the Golden Section exhibit natural proportions, the system can create order out of apparent chaos.[77] Hambidge's theory received mixed reviews. Many artists of the period rejected dynamic symmetry, while others, such as George Bellows, in an era of modernism, embraced the system's mathematical precision and rationality of composition.[78]

The Golden Section had been an object of study in American experimental psychology since the late nineteenth century, in empirical studies of aesthetic preference.[79] In 1871, Gustav Fechner had published *Zür experimentalen Aesthetik*, which called for a scientific approach to the psychology of aesthetic preference, and sparked a long tradition in psychological studies of aesthetics in general and of the Golden Section in particular. Many psychologists engaged with studies of the Golden Section during the early decades of the twentieth century, particularly within the new branch of applied psychology on the rise: the psychology of advertising.[80] By the late 1920s, with improvements in technologies of imagery, illustration was more prominent than the printed word in advertising.[81]

McCulloch's choice of subject for his B.A. thesis (written with "Professor Anderson") was not unusual or terribly original. He wasn't concerned with *why* the Golden Section was preferred by his subjects, but set out simply to find out if it was in fact the case. McCulloch designed a psychological experiment to test this assertion.[82] McCulloch presented various images on cards to his experimental subjects (twenty-four fellow Yale students)—representing root rectangles, the Golden Section, and images that displayed no dynamic symmetry. Based on their preferences, McCulloch concluded that there was indeed a preference for and ability to spot figures that exhibit dynamic symmetry.[83] Likely, McCulloch's choice of project was practical and dictated by the circumstances of his time at Yale and interests in psychology. However, it is plausible that he was drawn to the ways that Hambidge's theory of dynamic symmetry rationalized the natural world, brought order to it and to the processes of sense perception and aesthetic preference. McCulloch's continued interest in the Golden Section is revealing—it points to his long-standing fascination with mathematics and its intersection with art and perception, as well as the continual pull between the abstract world of numbers and the practical world of experiment.

## 2.6   Columbia: The Experimentalist

McCulloch's bio in the 1921 Yale yearbook states that upon graduation from Yale he planned to take up psychiatry.[84] Yet he entered a Master's program at Columbia and completed a degree in psychology in 1923, with a thesis that extended his work on dynamic symmetry. His supervisor here was applied psychologist Albert T. Poffenberger,[85] who had worked with Harry L. Hollingsworth to develop psychological tests for the armed forces during World War I, and had collaborated with him to produce an influential textbook in applied psychology, published first in 1917.[86] Poffenberger "championed" the used of illustration in advertising, in which aesthetics played a central role, in his 1925 textbook *Psychology of Advertising*.[87]

Psychology at Columbia epitomized the expansion of the discipline during the 1920s, and was beginning to enter a "golden age" dominated by the experimental-statistical approach of James McKeen Cattell.[88] In an era of enthusiasm for science and its potential for social control,[89] Cattell envisioned psychology as a natural science. Columbia had a prominent place in post–World War I American psychology, with Cattell's students Robert S. Woodworth and Edward L. Thorndike on faculty. Thus, the kind of psychology McCulloch was exposed to while at Columbia was firmly split from philosophy, although Columbia's department was rather eclectic.[90] In its applied psychology, there were no strong affiliations with clinical psychology or business, and most of what was practiced was quite theoretical and experimental, rather than practical.[91]

McCulloch's choice to attend Columbia and do a Master's degree in psychology, based on his B.A. thesis, may have reflected his original intention to study psychiatry following his undergraduate work, which would require medical school. McCulloch's undergraduate education only included a maximum of two hours of laboratory instruction in biology per week, not enough by American medical school standards by 1921. As the first two decades of the twentieth century witnessed increasing medical specialization and reform of medical education, medicine became increasingly based on the sciences.[92] During his Master's degree, McCulloch took courses in inorganic and organic chemistry as well as physics, philosophy, and psychology. By the end of his study he had clocked in over 300 hours of laboratory work.[93]

While a student at Columbia, McCulloch attended a party in New York City, where he met his future wife Rook (Ruth) Metzger McCulloch (1899–1991). Rook noticed a "tall young man who would go up to a group of people, listen to what they were saying, and then take the opposite view and convince them that he was right—then he would move to another group and do the same."[94] Rook had been born in New York City, daughter of David and Mina Metzger.[95] Although her parents were born Jewish, her mother was not religious, and Rook was agnostic, yet not antireligious. Rook and Warren were married on May 29, 1924. Their eldest daughter Katherine (Taffy) was born in 1926, adopted son George in 1927, David in 1929, and Mary Jean in 1930. During McCulloch's time at Columbia, the family divided their time between their home in Brooklyn and a summer residence in Sparta.

Following his Master's degree, McCulloch left the world of psychology forever and entered the world of laboratory medicine. He received his M.D. in 1927 from Columbia University, and was awarded the Certificate of the National Board of Medical Examiners in 1930. Columbia's medical school was well established: the College of Physicians and Surgeons was the first institution in the North American colonies to confer medical degrees.[96] McCulloch's first two years of medical school included courses in anatomy, histology, physiology, psychopathology, and biochemistry, both lecture and laboratory instruction. The third year was devoted to both theoretical and clinical instruction in medical specialties such as surgery and neurology, with fourth year focused on clinical instruction in hospital wards, case histories, diagnoses and prognoses.[97] Based on his medical school transcript, McCulloch was not a star pupil—during his first and second years he did not attain more than a B in any subject.[98] His third and fourth year grades were similar—all Bs except for an A in public health.[99] McCulloch did excel in physiology, and had exposure to both the laboratory and clinical contexts of brain research. Clinically speaking, he was fascinated by patients with injuries of the peripheral nerves, spine, and skull. According to his own reflections, McCulloch entered medical school "with the avowed intent of learning enough physiology of man to understand how brains work."[100] Within this world, the brain and mind were no longer objects of philosophical or psychological interest, but medical objects that could suffer from injury, disease, and disorder.

Although McCulloch's recollections suggest a continuous thread uniting his philosophical-psychological and medical academic pursuits, in fact as a student of medicine during the 1920s, McCulloch could not help but be swept into the world of the laboratory. As Harry Marks and numerous others have shown, laboratory science rose in the context of early twentieth-century American medicine as the basis of therapeutic techniques.[101] By the time McCulloch was in medical school, neurology—encompassing the distinct practices of clinical investigation, laboratory experimentation, and neurosurgery—was an established medical specialty.[102] Since the formation of the American Neurological Association in 1875, neurologists—some of whom worked in a more psychiatric tradition—became "scientized" and many centered on the study of diseases of the brain and nervous system. Experimental neurologists had strong ties to neurophysiology, and many based their practices on what John Fulton called "dynamic neurology" which was grounded in physiology and the work of British neurologist John Hughlings Jackson.[103] Hughlings Jackson had accepted that sensory and motor functions are localized within specialized areas of the cortex, yet believed that complex thinking and behavior must be "assembled" from simpler components involving separate areas of the brain.[104] For experimental neurologists who were dealing with conditions such as epilepsy, aphasia (loss of speech), and apraxia (loss of purposeful movement), this framework of localization could help confirm diagnosis in cases of brain damage. By the late 1920s, experimental studies of epilepsy had moved to the causes of seizures and had moved away from the use of human subjects toward laboratory animals.[105]

McCulloch's first forays into brain research were under the guidance of two titans of experimental neurology in New York City: Charles A. Elsberg, an influential neurosurgeon who had helped found the New York

Neurological Institute and worked to develop neurosurgery on a scientific basis, and epilepsy specialist Frank Henry Pike, professor in the medical school at Columbia. Both Elsberg and Pike advocated laboratory medicine, and both approached experimental neurology via the problem of localization.[106] Motivated by clinical considerations, they aimed to reproduce clonic seizures (involving frequent muscular spasms) and tonic seizures (more severe episodes with muscle stiffening, rolling back of eyes, and loss of consciousness) in the laboratory setting, working with the cat as an experimental animal.

For producing convulsions in an intact animal, Elsberg, Pike, and McCulloch injected increasing concentrations of absinthe intravenously, over the course of about two hours, and closely observed the physiological and convulsive effects in the animal, which began lying on its left side: pupils dilating, slight twitches, and nystagmus (dancing eye movement).[107] Initial lower doses led to clonic convulsions and larger doses led to tonic responses and eventually death.[108] McCulloch and his colleagues used lesions—anatomical ablation of parts of the cortex, particularly in the motor area—to study the effects of injury. They found that clonic convulsions arose from the motor cortex when intact and tonic convulsions arose from the lower motor mechanisms just after injury to the motor area of the cortex. They presented their research as having broad clinical and biological implications, as their study fed directly into the problem of the functional organization of the central nervous system as a whole.[109]

In 1928, McCulloch continued his neurology work at Bellevue Hospital in New York City with neurologist Robert Foster Kennedy, first as an intern then as a resident on Kennedy's neurological service. America's oldest public hospital, Bellevue was part of a long tradition of medical care in the United States that saw a sharp distinction between private or "voluntary" hospitals, in high abundance, and public hospitals, which were less common.[110] By the late 1920s, Bellevue had undergone the transformation that most hospitals in the United States had experienced: no longer simply shelters for the poor, hospitals were now sites of scientific medicine. McCulloch eventually collaborated with fellow intern Samuel Bernard Wortis in the Laboratory of Experimental Neurology at Bellevue,[111] and this work too dealt with the relations between head injuries and convulsive disorders. McCulloch and Wortis performed experiments on cats, producing injuries in three categories (brain lacerations, blood in subarachnoid spaces, and skull fractures) and observing their effects. They determined that brain laceration rendered animals more sensitive to a common standardized convulsant, camphor monobromide.[112] Although his work on epilepsy was situated in the experimental neurology laboratory, a variety of medical specialists engaged with this condition during the 1920s, moving across "long-standing intellectual and professional boundaries … in a number of major universities and hospitals. …"[113] Indeed, the 1920s and 1930s were characterized by an inter- and multidisciplinary ease of movement and collaboration between medical specialties.[114]

In 1928–1929, McCulloch and Rook purchased an old farm in Old Lyme, Connecticut, originally owned by Richard M. Champlin (who had died in 1871). They acquired it at a good price given the farm was in such a derelict state.[115] Rook initially turned the farm into Camp Aladdin, a haven for disadvantaged youth from Brooklyn. Rook and McCulloch managed to successfully broaden the outlook of individuals at the camp—it was a sort of social experiment where the kids made the rules—connecting them with artists and naturalists and introducing them to more practical skills such as carpentry.[116] This openness and generosity was shared by both Rook and Warren—the former having always been active helping others. Once the Depression hit, the camp closed, but Rook and McCulloch decided to make it a working farm. Ultimately, the farm became a refuge for the McCulloch family and for the many friends and guests they invited to stay—sometimes for extended periods of time.

When they had bought the land, the only buildings were the original farmhouse—which Rook and McCulloch restored—and a decaying barn.[117] Apart from the time he spent in the neurology lab, McCulloch set to work building up the farm. The lakes were dammed, and McCulloch both designed and engineered a new barn and a large stone garage, which he built with Rook's and David's help. The couple fenced and cleared the land, and Rook essentially became a farmer, McCulloch taking part when not at work. McCulloch's children paint a picture of a very practical, down-to-earth man:

I keep thinking of a Warren I knew that not many others saw. … He really knew tools. He knew what kind of wood had what properties that were needed for a job—he knew how to work rocks for building a masonry wall. He knew how to work metal in a forge and an anvil. … When he wanted to build a dam, I have seen his hydraulic calculations for the spillways. … His architectural drawings show lines he drew to fix the symmetry of the structure. Always symmetry—his early interest in the Golden Mean, and his long term interest in symmetry in nature and man's creations.[118]

This ease in both practical and theoretical matters helps us make sense of McCulloch's serial transitions between disciplines. From the worlds of experimental psychology and neurology, McCulloch soon moved from a laboratory to a clinical setting, a move that was in no way unusual given how fluid relations were between the sciences of the brain and mind during this period.

## 2.6   Psychiatry, von Domarus, and the Logic of the Diseased Mind

In 1932, in the depth of the Depression, McCulloch began working at the Rockland State Hospital for the Insane in Orangeburg, New York, about twenty miles from New York City. He worked in the Admission Service, in which patients would be treated for about nine months before transferred to a longer-term treatment ward.[119] Encouraging the notion of a "therapeutic suburb," Rockland was far removed, both literally and figuratively, from the urban context of Bellevue. While patients at Rockland enjoyed isolation, as a state hospital, Rockland dealt with more extreme psychiatric cases. The hospital was built to meet an increasing demand for space for psychiatric patients.[120] While at Rockland, McCulloch's scientific pursuits took a more clinical turn. While his disciplinary identity shifted from experimental neurology to clinical psychiatry, he was drawn to a brand of psychiatry that was less concerned with the realities of treating patients and more with the logic and underlying causes of mental illness.

The landscape of brain and mind sciences during 1920s was defined by eclecticism and a complexity of professional identity that characterized the relations between psychiatrists, neurologists, and neuropsychiatrists.[121] Overall, few generalizations about the relations between professional groups are possible.[122] Yet it is safe to say that in his move from experimental neurology to clinical psychiatry, McCulloch traversed a contentious if not fluid disciplinary boundary. Neurologists and psychiatrists were both rivals and collaborators.[123] While historically American neurologists and asylum superintendents had a relationship of professional competition—turning on questions about the relations between science and medicine and the nature of medical expertise[124]—by the 1930s, there was more fluidity between the worlds of the asylums and mental hospitals (the domains of alienists and superintendents) on the one hand and of public hospitals and private practice (the domain of neurologists) on the other. Since the end of the nineteenth century, American psychiatry had struggled for legitimacy and professional autonomy.[125] In the process, new standards of practice emerged in attempts to ground psychiatric practice in science and medicine. Especially following World War I, psychiatry enjoyed elevated status as practitioners worked less as asylum superintendents and more as physicians and researchers in hospitals, universities, and in private practice.[126]

Early twentieth-century psychiatry was characterized as a mix of clinical, laboratory, and theoretical practices devoted to making sense of the diseased mind.[127] Some searched for the physical and biological bases of mental illness, others wanted to ground psychiatry in physiology and psychology. Psychiatrists working within the framework of mental hygiene drew together mental illness with broad social and political goals and the social concerns of early twentieth-century America, particularly the eugenics movement, the fallout from World War I, and eventually the Depression of the 1930s.[128] The National Committee of Mental Hygiene, formed in 1909 in New York City, aimed to address mental illness as a social problem, through prevention, education, and clinical expansion in psychiatry. Others in the American context pushed for the development of new psychodynamic therapies, many of whom were influenced in varying ways by the work of Sigmund Freud.

Since 1909, when Freud delivered his famous lectures at Clark University, psychoanalysis had been taken up in the American context in complex ways.[129] While psychoanalytic practice came under attack by the medical establishment in America immediately following the Great War (an attack indicated by the founding of the journal *Archives of Neurology and Psychiatry* in 1919 to represent an alternative to the *Journal of Nervous and Mental Disease*, which was edited by an analyst, Smith Ely Jelliffe), artists, intellectuals, and laypeople welcomed Freudian ideas, and eventually professionals followed suit.[130] In order to make sense of Freud's influence in both popular and professional contexts, one must understand psychoanalysis not only as a medical technique or practice but also as a new way of understanding the world and human nature.[131] Popularly, Freud's work was at the heart of the "new psychology" of the 1920s. Within psychiatry, Freud appealed to various groups; interestingly and paradoxically, his framework transcended the intellectual divide between practitioners who promoted the idea of insanity as an illness of the mind—and thus emphasized social or environmental causative factors—and somatic theories wherein mental illness is a disease of the brain, determined by biological factors including but not limited to heredity, physiology, and anatomy.[132]

Psychodynamic therapies also proliferated during the 1920s in various forms. Besides the work of Carl Jung, Pierre Janet, and Morton Prince, hypnotism, suggestion, persuasion, and of course psychoanalysis were all common.[133] "Eclectic" psychiatrists also absorbed some aspects of the psychoanalytic spirit (but operated outside of the movement). The most prominent example of this sort of support was Swiss-émigré Adolf Meyer, who has been described as the most influential American psychiatrist of the early twentieth century.[134] His "psychobiology" integrated both an emphasis on environment and life experiences with a basis in biology. Meyer's framework for understanding dementia praecox, for example, characterized the illness as a failure to adapt successfully to one's environment.[135] Although not a psychoanalyst, Meyer promoted Freud's ideas in his training of students.[136] Ultimately, Meyer's psychobiology overshadowed psychoanalysis in professional circles.[137] His framework of mental illness as social maladjustment shaped the mental hygiene movement and influenced the transformation of psychiatric care.[138] Meyer's orientation was toward pathological research. Constructing detailed life histories, he used living patients as objects of investigation, in the tradition of Emil Kraepelin.

In the face of all of these various trends in interwar American psychiatry, McCulloch did not follow the path of a strict clinician, nor did he evolve into a strictly somatic or dynamically oriented practitioner. The eventual psychiatric perspective he adopted brought him squarely back into the realm of scientific philosophy. While at Rockland, he encountered two figures who were to serve as models for how scientific, medical, and philosophical pursuits could be integrated. The first was German-émigré psychiatrist Eilhard von Domarus, as well as von Domarus's Ph.D. supervisor at Yale, philosopher Filmer S. C. Northrop. Northrop fell within the trend of scientific philosophy discussed above, which continued to push metaphysics aside. In what Kuklick calls the "latest variant of pragmatism," associated with C. I. Lewis and his *Mind and the World-Order* (1929), studies of logic, scientific method, and epistemology came to the fore.[139] Scientists in turn were also taking an explicitly philosophical approach to their subject matter, and scientific and epistemological pursuits, for such scientists, were one and the same. Von Domarus typified this perspective—his approach to psychiatry was interdisciplinary partly because of his own background and training. He had received his M.D. from the University of Jena in 1922, and had trained with neurologist Hans Berger. Von Domarus also worked with Gestalt psychologists Wolfgang Köhler and Max Wertheimer and philosopher-psychologist Carl Stumpf in Berlin, as well as with Husserl and Heidegger in philosophy in Freiberg.

With a background that spanned the disciplines of neuroanatomy, psychiatry, and philosophy, von Domarus's ultimate aim was to connect scientific treatments of the brain to more philosophical conceptions of the mind and logical reasoning, and his thesis thus was presented as transcending the sciences of mind and matter. His work during the late 1920s while assistant physician at the Psychiatric and Nerve Clinic at the University of Bonn interpreted psychiatric conditions philosophically. These early publications concerned the absence of logical thinking in cases of schizophrenia, as well as errors in thinking that defined the behavior of manics, depressives, and those deemed "feeble-minded."[140] In 1928, von Domarus emigrated to the United States, and he received a Ph.D. from Yale in 1930, with a thesis entitled "The Logical Structure of Mind: An Inquiry into the Philosophical Foundations of Psychology and Psychiatry."[141] His supervisor was Northrop and he formally acknowledged help from Warren McCulloch.[142] The thesis, originally written in a mixture of Greek, Latin, German, and English (McCulloch assisted in translation) was firmly grounded in the tradition of German psychiatry and philosophy, and essentially pursued a history of scientific philosophy as well as a theory of knowing.

Von Domarus's framework for making sense of mental disease and conditions of the mind was imbued with a preoccupation with logic and foundations, and the question of what makes psychiatry distinct as a science. In a preface to the published version of his Ph.D. thesis, finished in 1934, von Domarus described McCulloch's interests, which clearly aligned with his own, especially McCulloch's fascination with the neurophysiological processes behind "anticipation, purpose, thought, and speech and in the product of these processes; specifically, in the propositions of fact and sentiment. The central problem for him is how organisms may be affected so that new propositions are induced."[143] This clearly was not conventional psychiatry. Chapter 1 of the thesis presented critiques of existing frameworks for making sense of the brain and mind. Von Domarus criticized the psychologies of J. B. Watson and Pavlov for neglecting discussions of the mind; he also at the same time attacked Freud's "dynamic psychology" for its imprecision and for its mode of treatment, which von Domarus described as "punitive and demonological."[144] Localizationists such as John Hughlings Jackson and Charles Pike seemed to have the most promise, for von Domarus, as they moved beyond the simple reflex model (which leaves most of the nervous system out of the picture) and attempt to "think through the function of the nervous system without resort to inhibition, shock, vicarious assumptions of function or hypotheses *ad hoc*."[145] In 1939, von Domarus presented to an English-speaking audience his ideas on logic and schizophrenia, part of a general study of the roles of language and thought in psychiatry.[146] After providing examples of errors in logic on the part of schizophrenics (e.g., feeling that a saint, a cigar package, and sexual life were identical by virtue of all of them being "encircled"—by a halo, a band, and a sexual glance), von Domarus concluded that schizophrenic reasoning is "paralogical," concluding identity based on similar properties.[147]

Describing his thesis to McCulloch in 1940, von Domarus admitted, perhaps understatedly, that the language was so mathematical that only one person "out of a hundred prospective readers ... would bother to attempt to study the thesis. This," von Domarus continued, "is a technical difficulty."[148] Furthermore, von Domarus told McCulloch, to properly appreciate the thesis, one would need background knowledge in sciences of the "inanimate, the animate, the mental, and the social." This along with other difficulties in accessibility—including knowledge of philosophy—von Domarus imagined would severely restrict the potential audience for the book.[149]

Von Domarus's supervisor at Yale was Northrop, and the former's philosophical views on psychiatry were infused with Northrop's discussions on the mingling of science and philosophy.[150] Northrop received his Ph.D. from Harvard in 1924, with the thesis "The Problem of Organization in Biology." In 1923, he gained a faculty position at Yale, where he remained for nearly forty years. Described as a philosophical generalist,[151] Northrop was an expert in philosophy, science, anthropology, and law, and this interdisciplinary framework informed his approach to philosophical and scientific problems. As one biographer wrote, Northrop "used the scientific method as his philosopher's stone."[152] In an examination of important events in the history of science, Northrop emphasized the importance of *theory*, arguing that "... what made a science out of chemistry was not mere observation and experiment but Lavoisier's attention to theory. ..."[153] Northrop also argued for the importance of the methodology of physics for biological science, and believed that formal logic and mathematics played a key role in the historical development of any science.[154] As a science develops historically, Northrop argued, its ultimate state of maturity is achieved through the incorporation of formal, deductive methods:

The history of science shows that any empirical science in its normal healthy development begins with a more purely inductive emphasis, in which the empirical data of its subject matter are systematically gathered, and then comes to maturity with deductively formulated theory in which formal logic and mathematics play a most significant part.[155]

Northrop illustrated his argument with the example of physics, which in his words had an "inductive" or "natural history" phase from the ancient Greek period to the Middle Ages, and gained a solid basis in "deductively-formulated theory" with the mathematical work of Galileo and Newton. Biology, for Northrop, was at present struggling with this transition from inductive to deductive methods.

Northrop's arguments for formalization in biology were connected to his larger philosophical-scientific vision of "dissecting the given scientific theories which ... scientists have verified, to determine what concepts and principles are taken as primary or undefined."[156] Through this process, scientific theories across all branches of science, including physics and biology, could be reduced to a set of primary, foundational concepts, or "first principles." This was the goal of Northrop's philosophy of science. More emphasis on theory and mathematical formulation in biology would allow physics and biology to be integrated. In Northrop's words, "there can be no adequate biological or medical theory of the concrete individual until there is a verified theory of the inter-relation of the basic concepts of the sciences ... to possess such a theory is to possess an experimentally verified philosophy of science."[157]

The influence of Northrop and von Domarus on McCulloch's perspective was profound. As McCulloch later recalled, von Domarus was not only a "superb clinician" but a "well-trained philosopher," and had been "forced into neuropsychiatry by philosophic problems."[158] Indeed, McCulloch called von Domarus the "great philosophic student of psychiatry" and enthusiastically emulated the former's mode of approaching psychiatric knowledge. During his work at Rockland with von Domarus, McCulloch began to understand cases of schizophrenia and psychopathia not from a traditional clinical perspective of the neurologist but rather as von Domarus had understood them, within a philosophical framework shaped by the writings of Bertrand Russell, Alfred North Whitehead, and Northrop.

## 2.7 Conclusions

By 1934, McCulloch had ten years of experience in the worlds of laboratory and clinical medicine. However, his association with von Domarus at Rockland renewed his philosophical interests and shaped his perspective on studies of the mind and brain. While his studies as an undergraduate and Master's student concerned human perception and processes of the mind, by moving into medical school, he entered a different world. McCulloch might have retrospectively fashioned himself as a philosopher interested in the problem of knowledge, and claimed this interest motivated his academic travels during the 1920s and early 1930s. Yet given the wider contexts in which philosophy, psychology, neurology, and psychiatry were developing during this period, McCulloch seemed driven more by a search for an underlying order to the natural world and to the human psyche in particular. Thus McCulloch's disciplinary trajectories and choices during this period cannot simply be attributed to an isolated philosophical interest in the problem of knowledge. Even early his career, if McCulloch was able to see past disciplinary frameworks and boundaries and was comfortable transcending such boundaries, this was motivated by his search for rationality in a world left irrational by the effects of war, a world that no longer made any sense.

McCulloch's transcending spirit was more than disciplinary. He performed various roles during this period—philosophy student, class wit, naval officer, modernist poet, experimental psychologist, husband, father, surveyor, farmer, experimental neurologist, psychiatry resident. In each role, in both his personal and professional worlds, McCulloch balanced reverence and iconoclasm, a panache for both the formal and the hands-on.

His choice of the sonnet as a poetic form placed him as part of a long line of traditionalists, yet the themes he tackled in his poetry were for the most part thoroughly modern. And while many of his poems were imbued with solemnity and longing for a "lost world," his persona as a college student was playful and irreverent. Scientifically, McCulloch was at home in the world of the laboratory as well as the formal world of theories of perception and abstract foundational discussions in psychiatry.

Given the permeability of disciplinary boundaries during this period, McCulloch was very much of his time. This made his travels easy and quite conventional. While traversing boundaries did not make him a nonconformist, his ease in different worlds of activity did.

# 3 The Neurophysiologist

## 3.1 Introduction

In 1934, in the depths of the Depression, Warren McCulloch left New York City and the world of clinical psychiatry, brought Rook and their four children back to the farm in Old Lyme to live full-time, and returned to Yale as a Sterling Fellow in the laboratory of Dutch-born neurophysiologist Johannes G. Dusser de Barenne. By then, American neurophysiology had experienced dramatic institutional expansion, fuelled in part by concerted efforts from various corners for the "scientization" of psychiatry. In a culture of fads, quackery, and popular enthusiasm for Freud, the psychiatry establishment, while still fragmented and diverse, aimed to ground psychiatry in basic laboratory research. Broadly speaking, diseases of the mind were increasingly understood to have an organic basis, and knowledge of the brain's basic physiology was needed to combat "unscientific" approaches to the mind. Within the walls of the neurophysiology laboratory, the brain as a scientific object began to gain a new valence beyond its traditional status as an object of clinical investigation. The Rockefeller Foundation poured vast amounts of money into building up neurophysiology laboratories and scientific studies of the brain, in an effort to "scientize" treatments of mental illness and nervous disorders. At Yale, Dusser de Barenne and McCulloch probed the brain not for clues about the source of epileptic convulsions, the effects of head injuries, or the physical bases of neuroses, but rather for knowledge of its workings in a state of health.

Simultaneously, this period also saw the continued expansion of scientific philosophy and interactions with like-minded scientists and philosophers, and Yale epitomized this development. As McCulloch moved from the clinic back to the laboratory, he also began to participate in discussions at Yale in a community of academics interested in the logical foundations of the sciences and scientific method. By the end of the 1930s, McCulloch began to articulate questions that transcended the divides between science, medicine, and philosophy. At this point in his life, McCulloch had dual identities—that of a laboratory neurophysiologist and occasional participant in a group concerned with epistemology and the logical foundations of the human sciences. Yet when viewed from the perspective of patronage, both practices reflect general trends toward the rationalization of the human mind and of the methods used to study it, be it within the walls of the laboratory or via more "interstitial" practices.[1]

McCulloch straddled the division between two scientific worlds during this period. While this was by no means unusual, his particular identity as a neurophysiologist was more than simply a reflection of his institutional contexts. While he may have fashioned himself as a scientist "forced into neurophysiology" by philosophical interests, the relations between these two pursuits was more nuanced and complex. McCulloch was intensely drawn to practical knowledge—hands-on experiences—whether it be on the farm or in the laboratory. Yet he continued to be fascinated by the abstract world of foundations, logic, and mathematics. His identity as a neurophysiologist was performative—not simply a reflection of context but idiosyncratic. Professionally he was a neurophysiologist, yet performatively, an experimental epistemologist: forging a scientific theory of knowledge based on the brain's functional organization.

## 3.2   Freud, Fads, and the Fragmentary Nature of American Psychiatry: Efforts toward Legitimacy

The Great Depression in America had sparked a flurry of popular literature for "self-improvement," as the economic well-being of the nation was tied to individual health.[2] For popular writers who worked in a eugenic framework that stressed degeneration and the fundamental irrationality of the human mind, self-control and efficiency were paramount. The 1920s had been an active time for new religions, new spirituality, and alternative healing systems.[3] Psychology had become popularized through ubiquitous "how-to-become" literature and other forms of self-help books. Much of this work amounted to a dilution of Freud, and lay analysts were everywhere.[4] John B. Watson's behaviorism—which reduced human behavior

**Figure 3.1**
Warren McCulloch in 1938. Courtesy the McCulloch Family.

to conditioned responses governed by instincts—also contributed to the "new psychology" of the era.[5] While secular works had replaced many of the old religious-based writings,[6] superstition, folk beliefs, and occultism also flourished.[7]

In the face of all this, by the early 1930s, academic psychiatry was struggling to achieve professional legitimacy, and in the process, bring harmony to a fragmented, eclectic discipline. Neuropsychiatrists refashioned themselves and their practice as addressing "all the factors that disturbed the path of social progress."[8] Psychiatry during this period was mostly preoccupied with the small deviations from normal—what Elizabeth Lunbeck has called the "psychiatry of everyday life."[9] In part this was the result of a move by the early 1930s from the asylum culture to hospital culture as the main focus of psychiatric practice. Ultimately, Adolf Meyer's focus on "social maladjustment" became central to the eventual psychiatric renaissance of the period, as state hospitals also moved from being strictly custodial to being sites of clinical research.[10] As we saw in chapter 2, Meyer's "psychobiology" sought to reconcile an individual's biology with their life experiences.[11] His efforts can be read as an attempt to bring objectivity to the field of psychiatry by reinforcing the discipline from within: although psychobiology implied an organic basis to mental illnesses, Meyer wished to demarcate their study and treatment as solely within the purview of psychiatry, not neurology or neurophysiology per se.[12] Yet Meyer's psychobiology was an attempt to transcend the divide between the natural and human sciences.[13]

Indeed, psychiatry's transformation had much to do with its relations with the natural sciences. In 1932, with support from the Carnegie Corporation, the Committee on Psychiatric Investigations of the National Research Council began a study on "The Problems of Mental Disorder," chaired by psychologist Madison Bentley and cytologist E. V. Cowdry. They aimed to bring together researchers from various disciplines to discuss methods for "advancing our understanding and control of the mental disorders."[14] One facet of this project involved soliciting help from the natural sciences to "suggest methods and issues of research in the sciences."[15] In an era where "the family, the neighbourhood, and the state were all desperate," disciplinary strength was needed to educate and equip psychiatric practitioners.[16] Following contributions by members from the core psychiatric disciplines, the bulk of the published volume included essays from the "supporting sciences," e.g., anatomy, physiology, electrophysiology, genetics, nutrition, and endocrinology, in an effort to bolster psychiatry via interdisciplinary collaboration.[17]

Another facet of reform efforts in psychiatry dealt with its relation to medicine and the extent to which it was scientific. At the start of the New Deal era, psychiatry did not have a firm place in medical school curricula, and was viewed by many to be lagging far behind knowledge in other areas of medicine.[18] As a report of the Commission on Medical Education noted in 1932,

Our meager understanding of the underlying factors which produce many of the mental, emotional, and functional disorders is comparable to that which existed in regard to general medicine fifty years ago before the introduction of accurate, measurable criteria of study, which transformed medicine and surgery from mysticism, empiricism, and guesswork to what in many instances now approaches scientific certainty.

No subdivision of medicine has received so much publicity and propaganda by untrained workers as the nervous and mental disorders, in which our present knowledge is still fragmentary and uncertain. Probably in no field of health work is there so much dangerous faddism. This is unfortunate because the problem is one of the most important health, educational, and public questions of our time.[19]

By the early 1930s, efforts to strengthen relations between psychiatry and the natural sciences, and to raise the status of psychiatry in medicine increased. New links also emerged between psychiatry and neurosurgery.[20] Yet psychiatry was still a fragmented, heterogenous discipline defined by disharmony.

### 3.3 Scientizing Psychiatry: The Rockefeller Foundation's Science of Man Program

By the start of the 1930s, the laboratory had become the measure of what made medicine scientific. Grounding psychiatry in the laboratory sciences was one motivation of the Rockefeller Foundation (RF), whose efforts to build up psychiatry in the United States during the 1930s was a cornerstone of its program for a "Science of Man." In the early decades of the twentieth century, American philanthropic foundations began to share a more general enthusiasm for science—which, in the eyes of wealthy industrialists such as John D. Rockefeller (who had accumulated a vast fortune in oil), was the basis of the nation's industrial and economic strength.[21] Between 1900 and 1925, the RF became the largest supporter of medical research in the US.[22] This support amounted to more than simple enthusiasm for science—RF programs were directed toward the "improvement of mankind" and the value of scientific solutions to social problems.[23] As the work of Lily Kay has shown, broadly speaking, the Rockefeller Foundation agenda during the 1920s and 1930s was both a scientific and cultural enterprise, conceived within a framework "social control."[24] A synergy existed between Christian values and the Rockefeller Foundation's "science-based social mission," exemplified in the vision of Raymond B. Fosdick, originally a trustee of the foundation and later president.[25] Following a major reorganization of the Rockefeller Foundation in 1928,[26] the Foundation's funding priorities in medicine had shifted from medical education to an increasing emphasis on medical *research*.[27] The "Science of Man" agenda, which focused on problems of human behavior, emphasized the role of biological causation, experimentation, and quantification.[28]

Since the early 1930s, the Medical Sciences Division, steered by Alan Gregg from 1931 to 1951 and by Robert S. Morison between 1951 and 1959, had placed emphasis on the clinical fields of psychiatry and neurology in its support of brain research, as well as the related fields of neurophysiology and neuropathology.[29] Mental illness could be approached from a common framework whether the practitioner came from the private clinic, state hospital, or the laboratory bench.[30] Under Gregg's influence, psychiatry became a top priority in the RF agenda for the medical sciences.[31]

In 1930, David Edsall, Dean of Harvard's medical school, was asked by the RF to provide an assessment of the current state of psychiatry. He argued that physiology and laboratory methods (such as those of Ivan Pavlov, Walter B. Cannon, and Charles S. Sherrington) would rescue psychiatry from the dominance of "elusive and inexact methods of study and by speculative thought."[32] Russian physiologist Pavlov had performed laboratory studies of the conditioned reflex and demonstrated for many the worth of experimental physiology for understanding human behavior and the connection between body and mind. A contingent of American physiologists had used Pavlov's work as a framework for making sense of how the body and mind could be understood experimentally, and Pavlov's "physiology factory" became a model of scientific production.[33] Harvard's Cannon had done pioneering work in human physiology, and Sherrington's experimental and theoretical work on the central nervous system, which will be discussed below, shaped an entire generation of American and British neurophysiologists.

Furthermore, Edsall continued, the subject matter of psychiatry had necessitated a broad line of attack, which had not always been fruitful. His recommendation was to work toward establishing laboratories, so that men could be trained and so that the problems of psychiatry could be attacked by "exact" and "precise" methods.[34] Mental illness, according to Edsall, should be confronted by less speculative approaches and instead be modeled on more empirical disciplines. His key example, as suggested by his earlier allusion to the work of Pavlov and Sherrington, was brain physiology: "It is reasonable to think that many things will remain obscure until we know much more about the organ in which the mental processes take place."[35]

Edsall's perspective directly informed Gregg's vision for the Rockefeller Foundation's program in psychiatry, and he gave the laboratory a central role in shaping the foundation's program. Not only was Gregg informed by Edsall's emphasis on basic research, but he also moved Rockefeller Foundation policy in the direction of support for interdisciplinary, integrative collaborations between the sub-fields that comprised psychiatry.[36] The institutional programs that benefited from Gregg's program for the medical sciences were all interdisciplinary and committed to a unity of brain and mind.

In April 1933, RF officers directed their efforts toward the "general problem of human behavior" and called for coordination between the natural and medical sciences divisions.[37] Later that year, Gregg and Natural Sciences Division director Warren Weaver presented a combined program in psychiatry and the "sciences underlying the behavior of man."[38] The conceptual basis of the program was psychobiology, which in their vision encompassed psychiatry, psychology, and neurophysiology. Quantitative techniques should be applied to biological problems, and experimental work on the nervous system will be essential for understanding central problems in both psychology and psychiatry. Mind and body were merely seen as "separate aspects of a unitary self."[39] Gregg acknowledged that psychiatry as a discipline is largely clinical, yet neuroanatomy and neurophysiology were basic to a study of the diseased mind.[40]

### 3.4 Neurophysiological Practice: Cerebral Localization

Within the policy framework of the RF Science of Man agenda, neurophysiology as a discipline flourished. Laboratory studies lent scientific credibility to psychiatry, and the medical and cultural motivations for making psychiatry scientific allowed neurophysiology to thrive. Thanks to RF funding, neurophysiology was built up around instruments and material practices for generating knowledge about the brain and peripheral nervous system,[41] and graphic representation became a significant feature of this process.[42] Instruments were designed to amplify, measure, and record—in essence make visible—the electrical and functional activity of the nervous system.[43] The electroencephalograph, the cathode ray oscilloscope, the microelectrode, and the stereotaxic instrument defined laboratory practice at key institutions such as the University of Chicago and Washington University in St Louis. For example, the practice of electroencephalography (EEG), the recording of human "brain waves," first performed by Hans Berger in 1929, was a highly locally contingent practice that ultimately transformed brain research.[44] Initial EEG studies were on the human brain, but eventually the graphic technique of producing waves from the cortex of experimental animals became ubiquitous in attempts to understand the brain and its functioning. However, the EEG gave little direct information about the activity of single nerve cells in the brain or what areas of the brain were contributing to the recorded pattern.

Scientific practice in biomedicine involves a variety of tools, and such tools can modify disciplines and be central to their development.[45] As focus in American neurophysiology shifted to the central nervous system, theoretical-material tools also became key for organizing knowledge about the brain. The practice of cerebral localization, which involved matching mental, behavioral, and physiological functions to specific regions of the brain, was at the heart of American neurophysiology during the 1930s, and had both theoretical and material dimensions.[46] Brain localization was at once a material practice for investigating the activity of the brain itself, and a theoretical practice for gaining knowledge about the nature of the brain's link to behavior and how various parts of the brain were organized and worked together. Susan Leigh Star has argued that these two facets of localization are so closely linked they can be considered indistinguishable. Practitioners studying the brain aimed to make sense of how this complex mass of neurons could be structurally and physiologically responsible for countless sensory and motor functions of the living organism.

Localization as a practice has a long history. As both a basic and clinical practice, localization of brain function not only dealt with the nature of the brain but had relevance for philosophy, theology, physiology, and psychology. Star has shown a myriad of positions and revisions in localizationist theory across the divide between the laboratory and the clinic during the late nineteenth and early twentieth centuries in Britain. Accounting for these theories sociologically, she argues that the success of localizationists was largely based on commitments to "ways of working."[47] British localizationists came from a variety of disciplines that spanned the basic and clinical realms, such as John Hughlings Jackson, David Ferrier, Victor Horsley, and eventually, Charles Scott Sherrington. Triangulation defined research practice here, whereby evidence from different research domains was combined to study localization.[48] Overall, theories about the localization of function in the brain were "distributed" over time and place, and as a result, uncertainties in local contexts—autopsies, experiments, surgeries—could be overcome.

Of course, such theories had implications for philosophy and the mind-brain problem. By 1906, Sherrington's integrative framework of functional organization—a descendent of Hughlings Jackson's understanding of the brain—was dominant as a theoretical tool for making sense of the brain's structure and function. Although practitioners such as Hughlings Jackson and Sherrington emphasized the physical realm, their psycho-physical parallelism—where brain and mind operate "in tandem but as completely separate and sovereign realms"—resulted in what Star calls a "philosophical deadlock" where the problem of their interaction was avoided.[49]

The American context of localization was quite distinct by comparison, although most practitioners inherited a tradition of theory and practice from the British context. American neurophysiologists had acquired a distinct disciplinary identity, with only partial connections to the realm of the clinic.[50] As we saw in chapter 2, as an experimental neurologist at Bellevue during the late 1920s, McCulloch had worked within a Jacksonian framework, although Sherrington's work was most widely influential. The most eminent physiologist of his generation, Sherrington's major synthesis, *The Integrative Action of the Nervous System* (1906), presented the nervous system as a structurally and functionally integrated system of nerve cells. Nervous reaction, for Sherrington, integrates a multicellular animal, and "welds it together from its components, and constitutes it from a mere collection of organs an animal individual."[51] The unit mechanism of the integrative nervous system was the reflex arc, which in and of itself Sherrington viewed as an integrative action. Working with anthropoid apes such as the chimpanzee and gorilla, and using techniques such as lesion and stimulation, Sherrington and his associates aimed to correlate changes in motor cortex with motor activity. Assuming a correlation between physical states and mental events, Sherrington argued that any understanding of psychological activity was best approached through physiological study of the brain.

Within the diffusionist-localizationist spectrum of positions on brain function, Sherrington occupied a unique position. While his integrative model of brain function bore a striking similarity to the diffusionist model of Mauritian physiologist Charles-Édouard Brown-Séquard, Sherrington was interpreted as a localizationist.[52] This was not philosophy of mind in the strict sense, which would address more precisely the question of, for example, dualism or monism, and within the dualist framework how the brain and mind interact (e.g., psychophysical parallelism). Rather, this concerned a scientist's *theoretical position* on *brain organization*. Others rejected strict localization. For example, American neuropsychologist Karl Lashley rejected the idea of strict localization of function in the brain on the basis of evidence which showed the brain to function much more dynamically, particularly in learning.[53]

Sherrington was mentor or collaborator to a large number of prominent physiologists in the English-speaking world of the 1930s and 1940s, many of whom ultimately played a foundational role in American neurophysiology, particularly Yale neurosurgeon Harvey Cushing and Yale neurophysiologist John F. Fulton, and in 1924, Dusser de Barenne.[54] All continued Sherrington's integrative tradition, which involved not only material practice but also the more cognitive and theoretical tools that defined American neurophysiology.[55]

## 3.5  Yale University

Yale University became an epicenter of the disciplinary dynamics that defined brain research during the 1930s and 1940s. McCulloch's return to Yale in 1934 took place at a time when studies of the brain were flourishing in an institutional climate that favored interdisciplinarity. Milton Charles Winternitz became Dean of Medicine in 1920 and headed a reorganization of the school. A pathologist trained at Johns Hopkins, Winternitz restructured the medical school units as university departments and sought institutional unity.[56] A symbol of this transformation was Yale's new medical school building, the Sterling Hall of Medicine, financed in part by the Rockefeller Foundation's General Education Board, dedicated in 1925 and funded by the John W. Sterling Trustees. It achieved physically the unity that Winternitz strove for, bringing together in one place departments that previously had been disconnected. The building had departmental space, laboratory space, and space for a library and lecture theater.

The story of the Rockefeller Foundation and Yale's medical school must be understood from the perspective of the RF's "Science of Man" program.[57] A key facet of the medical and human sciences at Yale was the Institute of Human Relations (IHR), which was founded in 1929 to bring together disciplines from all over the campus into a unified and organized institutional structure and to integrate "all the teaching and research at Yale that pertained to the study of man."[58] The IHR was placed firmly *within* the School of Medicine and one of the driving forces behind the IHR was an emphasis on disciplinary unification, in the face of a new emerging image of the mind as non-rational and non-autonomous, derived from Freudian psychoanalysis.[59] Psychologists strove to make their discipline more scientific, and the IHR aimed to control human behavior by mechanizing and rationalizing it. The IHR received over $4.5 million from the Rockefeller Foundation during its first ten years. Winternitz envisioned interaction between psychologists, psychiatrists, and neurologists, coordinated by the IHR, which, rather than being a department was a "voluntary association of scientists."[60]

The star of basic brain research at Yale's Medical School during the 1930s was John F. Fulton, and the revival of interest in neurophysiology in America during the 1930s owes much to Fulton's administrative and scientific talents. Fulton had worked with Sherrington at Oxford for six years during the 1920s on the physiology of muscular and reflex control,[61] and he played an important role in communicating Sherrington's ideas—particularly about integration—in the American context.[62] Fulton viewed neurophysiology as basic to the study of neurology and psychiatry,[63] and this was reflected in his practice of cerebral localization.

Fulton approached Alan Gregg early in 1933 with an assessment of the state of American neurology—as per Gregg's request.[64] Fulton emphasized that the most important work in neurology being done in the US at the time was what he called "dynamic neurology," based on neurophysiology. Fulton told Gregg that the only basis for sound neurological diagnosis and therapy was the analysis of neurophysiological mechanisms. This was neurology, Fulton said, in the tradition of Hughlings Jackson, Sherrington, and Rudolf Magnus.[65] Referring to a series of articles that had recently appeared in the American Medical Association's journal *Archives of Neurology and Psychiatry* on the training of the neurologist (an effort to address the shifting professional status of neurologists in light of increasing medical specialization), Fulton departed from a more traditional, "insular" perspective on neurology. Fulton believed basic knowledge in a clinical context was crucial: prognosis and therapy for the patient would be based on knowledge of the neurophysiological mechanisms involved in the neurological disturbance.

Fulton's philosophy of neurophysiology as key to the development of neurology and psychiatry fed into a request to the Rockefeller Foundation in June 1933 for funding for postgraduate training in neurophysiology, particularly for medical graduates who propose to take up neurology or psychiatry.[66] Fulton succeeded in securing Rockefeller funding, mainly for primates, equipment, and supplies for the use of postgraduates in neurophysiology who were at the Laboratory of Physiology.[67] Fulton's Department of Physiology at the start of the 1930s had three independent divisions: Fulton's Laboratory of Neurophysiology, Robert E. Yerkes's Laboratories of Comparative Psychobiology, and Johannes G. Dusser de Barenne's Laboratory of Neurophysiology.[68] McCulloch arrived in Dusser de Barenne's laboratory in 1934.

### 3.6   Dusser de Barenne: Background

In Dusser de Barenne, McCulloch could not have asked for a more apt mentor. Johannes Gregorius Dusser de Barenne [figure 3.2] was born in 1885 in Brielle, The Netherlands, and entered the University of Amsterdam, receiving his medical degree in 1909. His biographers, Fulton and Gerard, divided Dusser de Barenne's career into five distinct periods.[69] Between 1909 and 1911, Dusser de Barenne worked at the Laboratory of Physiology at the University of Amsterdam, investigating the effects of the local application of strychnine on the central nervous system, particularly spinal reflexes.[70] Strychnine, a highly toxic plant extract, was to become the central tool for Dusser de Barenne's investigations of the central nervous system throughout his career.[71] The second phase of his work, from September 1911 until the outbreak of the Great War, saw Dusser de Barenne working as psychiatrist at the Meerenberg Lunatic Asylum. During the third phase of Dusser de Barenne's career, during World War I, he served on active duty as a medical

**Figure 3.2**
J. G. Dusser de Barenne, 1936. Yale Laboratory, New Haven, Connecticut. From R.
McCulloch, *Collected Works of Warren S. McCulloch, Volume I* (Salinas, CA: Intersystems, 1989).

officer in the Dutch Army, stationed at Delft during the mobilization. Just
before the war ended, Dusser de Barenne accepted an appointment as Lecturer and Privat Dozent in the Departments of Pharmacology and Physiology at Utrecht University, and held this position until September 1930.
During this fourth phase his career, Dusser de Barenne worked closely with
the eminent German physiologist and pharmacologist Rudolf Magnus on
the physiology of posture.[72]

Magnus had early on been drawn to literature and philosophy, but ultimately studied medicine and specialized in physiology and pharmacology.
His work on physiology of posture began with his short collaboration with

Sherrington in Liverpool in 1908, where he studied reflex mechanisms. He also accepted a position as chair of pharmacology at Utrecht that same year, where he remained for the last eighteen years of his life. Magnus became well known for his work during this period on the factors controlling animal posture, particularly his joint studies on posture with A. de Kleijn, and in 1924 published a book based on their studies.[73]

Magnus's early interest in philosophy remained throughout his career. In 1906 he delivered a series of ten lectures on Goethe, and he brought a philosophical perspective to his work in physiology.[74] He had been well known for his theory of the "physiological *a priori*," which he described in his three Lane Medical Lectures at Stanford University in 1927.[75] Immanuel Kant's synthetic a priori, outlined in his *Critique of Pure Reason*, refers to the idea that there are a variety of elements given a priori that we are compelled to employ in our experiences of thinking and drawing conclusions about the world around us. Magnus argued that Kant's synthetic a priori is normally interpreted "philosophically-psychologically," that is, as an aspect of the psyche. However, Magnus argued that the a priori must also have a physiological basis. This was related to the notion that one does not come to sensory data as a "blank tablet," but rather brings a sort of relational structure within the nervous system to interpret sense data. The nature of sensory impressions is determined *a priori*, by the physiological sensory apparatus of the brain. The physiological mechanisms of our central nervous system allow us to make sense of the sensory observations that come to us from the outside world. Reflexes were interpreted the same way by Magnus, as contributing to posture and eye position, they contribute to acts of perception.[76] While there is no direct evidence for Dusser de Barenne pursuing these same overtly philosophical lines of inquiry, his work with Magnus certainly colored his perspective on the brain and his attempts to make sense of its functioning as an organized system responsible for sensory perception.

By 1916, Dusser de Barenne's attention had turned to localization of function in the cerebral cortex. He believed the method of stimulation had yielded no significant results, as he observed "an animal newly awakened from narcosis can give no information as to sensory disturbances produced by electrical stimulation of the cortex, even if any disturbances were thereby caused."[77] He felt that both "experimental" and "cytotectonic" methods (histological methods based on patterns of cell structure in the cortex) had not had not produced a firm basis for understanding localization of sensory functions.

In light of the fact that the local application of strychnine had been useful in Dusser de Barenne's investigations of the spinal cord, he decided to use the tool to investigate the cerebral cortex. He saw several advantages of using the strychnine method over the extirpation method. While he acknowledged the important contributions of studies employing the lesion method, Dusser de Barenne presented the strychnine method as superior for producing symptoms of excitation in the sensory area of the cortex, mainly because of its precision and intensity:

In successful experiments, when the cortex had been exposed as carefully as possible and without any unfavourable concomitant circumstances, I have never observed any disturbances in the animals when they have awakened from narcosis; they frequently jump down from the table, with all their movements precise and natural ... in spite of undoubted injury to the cortex ... the symptoms of excitation after the application of strychnine were clear and unmistakable in the same part of the body in which the signs of impairment had appeared.[78]

Thus, Dusser de Barenne concluded, the excitation produced by strychnine was intense enough to mask any effects from injury. Dusser de Barenne's early work using the method of "local strychnine poisioning" involved experiments on the cerebral cortex of the cat. In each procedure, the cat was first anaesthetized using chloroform and ether. The region of the cortex to be experimented on was then exposed a bone-cutting surgical instrument and bone forceps, and any excess cerebro-spinal fluid was absorbed by dabbing the surface of the cortex with cotton. A 1-percent strychnine solution, colored with toluidin blue, was then applied to the cortex using a tiny wad of cotton wool at the end of forceps, and any excess strychnine solution was removed. The resulting poisoned spot on the cortex was then seen as a small blue area of a few square millimeters. The cat's skin was then stitched back, to prevent cooling. According to Dusser de Barenne's description, the entire procedure could be performed within a span of ten to fifteen minutes. Following the cat's recovery from narcosis, one could then observe and compare symptoms when the sensory cortex was stychninized within a certain region of the cerebral cortex and outside this same region, observing disturbances in the cat such as paralysis and hypersensitivity. In the spring of 1924, Dusser de Barenne went to Sherrington's laboratory at Oxford, to study sensory symptoms through the application of strychnine to the cerebral cortex of rhesus (macaque) monkeys.[79] He refined his technique here and produced results that delimited the sensory cortex of the monkey, producing a classic study that delimited the major subdivisions in the sensory cortex—the leg, arm, and face. This study launched a major research program for Dusser de Barenne that defined his work during the 1930s, the most productive phase of his career.

In the spring of 1929, Winternitz visited Utrecht hoping to convince Dusser de Barenne to come to Yale. That summer the International Physiological Congress took place in Boston as well as the International Psychological Congress a week later in New Haven. Dusser de Barenne was at both meetings and eventually decided to come to Yale in September 1930, thus beginning the fifth phase of his career. Winternitz gave him the opportunity to establish a research laboratory in neurophysiology within Fulton's Department of Physiology,[80] and by summer 1931, Dusser de Barenne's lab moved into its permanent quarters, which included rooms for work on conditioned reflexes, a sound-proof room for acoustic investigations, one for chemical work, one for histological work, and one for electrophysiological experiments [figure 3.3]. Genial, friendly, and an admired teacher, Dusser de Barenne attracted a large group of students and assistants, including Percival Bailey, Leslie Nims, Arthur Ward, Craig Goodwin, Hugh W. Garol, and Warren McCulloch.

### 3.7   Mapping the Brain: Dusser de Barenne and McCulloch

McCulloch's early collaborations with Dusser de Barenne, funded mainly by Yale's Fluid Research Fund, essentially involved translation and extension of Sherrington's notion of functional organization.[81] They performed electrical stimulations of the motor cortex of the monkey, to study the phenomenon of cortical "extinction" or inactivation.[82] These stimulation experiments led Dusser de Barenne and McCulloch to conclude that extinction was a highly localized phenomenon, most probably connected to the pyramidal cell layer of the cortex. By 1936, Dusser de Barenne had introduced McCulloch to the strychnine method for localizing functions in the cerebral cortex.[83] Coupling the strychnine method with recording action potentials in the monkey using a cathode ray oscillograph, they established that there were functional boundaries between the main

**Figure 3.3**
Large general experiment room, Laboratory of Neurophysiology, Sterling Hall of
Medicine, Yale University, 1932. From J. G. Dusser de Barenne, *Methods and Problems
of Medical Education, No. 20* (New York: Division of Medical Education, Rockefeller
Foundation, 1932), pp. 25–26.

subdivisions of the sensory-motor cortex. Most interestingly, their work
revealed that there were *directed* functional relations between areas: for
example, they observed that if one strychninized region A, spikes were
recorded from region B, but if region B were strychninized in a separate
experiment, no spikes were recorded from region A. McCulloch's early
work with Dusser de Barenne confirmed the latter's earlier hypothesis that
complex functional relationships exist between different areas of the
cortex.[84]

During a highly productive period in the spring of 1938, Dusser
de Barenne and McCulloch published four papers together in the
newly established *Journal of Neurophysiology*, in which they applied the
strychnine method to the study of the sensory cortex of the monkey
(Macaca) and its relation to the thalamus.[85] Their results integrated anatomical and physiological analysis of the cortex, and confirmed the existence of functional relationships between different areas. In this way,

Dusser de Barenne's work went beyond straightforward localization and rather focused on the *functional organization* of the cortex: the influence of one cortical area upon another, and the interaction between different areas of the cerebral cortex [figure 3.4]. Working within a Sherringtonian framework, Dusser de Barenne was critical of "classical" localization theory, with its assumption of a "sharp, point to point, geometrical projection of the body on the cortex."[86] He viewed the functional organization of the cerebral cortex as complex and plastic, and in 1934 argued that "with regard to the cortical representation of the somatic functions, there is not one type of functional localization in the cortex, but more, perhaps as many as there are senses."[87]

Dusser de Barenne's and McCulloch's work on functional organization was made even more precise in their efforts to delimit neurons in the cerebral cortex, using the strychnine method in a procedure they called "chemical neuronography" or "strychnine neuronography."[88] Similar to the methods used in their earlier work, their goal here was to understand "communication" in the cortex—a hint of a general process that later occupied members of the cybernetics group—by deducing specific pathways of neural impulses. Building on their previous work, Dusser de Barenne and McCulloch aimed to correlate what they had found concerning functional boundaries in the cortex with its specific neuronal structure—to determine if, in the strychninized area, neurons *originate* that *end* in the area where electrical activity is recorded. Drawing on neuroanatomic evidence regarding the direction of neuronal connections in the cortex, Dusser de Barenne and McCulloch concluded that when they strychninized a particular region A of the cortex, and recorded "spikes" from region B, the neurons that were strychninized in region A indeed had an ending in region B. They argued that local strychninization, coupled with the recording of action potentials, was a powerful tool for delimiting the origins and endings of neurons in the central nervous system.

In terms of his scientific practice, Dusser de Barenne had an unyielding faith in experiment. As Fulton and Gerard noted with respect and admiration, he had an "utter intolerance for those who place the armchair ahead of the experimental table as a place for solving the problems of physiology—'never think, if you can experiment'."[89] Widely read in philosophy and art, among his heroes were French physiologist Claude Bernard (whose portrait stood on Dusser de Barenne's desk), Carl Ludwig,

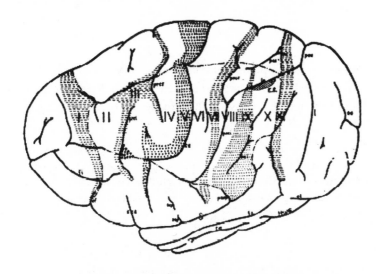

## FUNCTIONAL ORGANIZATION
## SENSORY ARM CORTEX

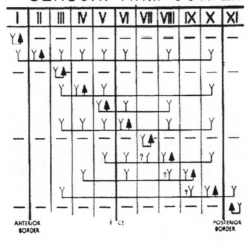

**Figure 3.4**

McCulloch and Dusser de Barenne on the functional organization of the cerebral cortex (1941). *Top*: The extent, location, and functional subdivisions of the sensory arm cortex into physiologically distinguishable bands, Nos. II–X, and of the immediately adjacent bands, Nos. I and XI, indicated on a composite drawing representing the arm area in the center of the field. The bands giving suppression, Nos. I, II, VII, and XI, are marked with horizontal lines. Small stipples indicate bands V and IX, large stipples mark the "dud" areas. *Bottom*: Diagrammatical representation of the directed functional (and anatomical) relations between the various cortical bands of the arm-subdivision of the sensory cortex [from original caption]. From J. G. Dusser de Barenne, Hugh W. Garol, and W. S. McCulloch "The "Motor" Cortex of the Chimpanzee," *Journal of Neurophysiology* 4, no. 4 (1941): 287–303.

and Rudolf Magnus—all staunch experimentalists.[90] McCulloch's recollections reinforced this picture of Dusser de Barenne as a relentless experimenter. As McCulloch recalled, Dusser de Barenne would insist on going on for 55 hours,

> repeating an observation every three minutes on a single point on a single hemisphere before he was satisfied. He was slow and uncertain of his own reasoning powers when the argument became abstract or mathematical and would insist that the question was either just words or could be reduced to a question of fact and settled by experiment at once. Thus, theory was compelled to be subservient to experiment at every step, and the step was taken then and there.[91]

In McCulloch's view, this approach helped Dusser de Barenne become a pioneer in the physiological study of the chemistry of the brain, and in investigations of the functional organization of the cerebral cortex.[92]

McCulloch worked hard during this period—six days a week, reportedly, spending all day Sunday sleeping.[93] He balanced his laboratory work with activities on his 600-acre farm. A big barn was built in 1938, and in 1940 a stone millhouse was erected, with the farm operating a dairy herd during World War II. While Rook had been a city dweller, by the end of the 1930s

**Figure 3.5**
Rook McCulloch with three of her children, Mary Jean, Taffy, and David, circa 1940. McCulloch Papers, Folder "Jean, Taffy, Rook, and David McCulloch, [ca. 1940]; 1 print," Box 4, Series VIII Photographs, WSM Papers, APS.

she had become a farmer, undaunted by the hard work such a role entails. McCulloch's daughter Mary Jean recalls that while on a ladder, McCulloch called down to Rook: "'Kitten, hand me that rock.'... He pointed to a fairly big one. She picked it up and climbed the ladder with it."[94]

Laboratory work was intense as well. McCulloch has described his years with Dusser de Barenne at Yale as exciting and full of discovery—"so much so that Dusser would slam me on the back and shout: 'We discover too much!'"[95] McCulloch collaborated with Dusser de Barenne for six years, and they ultimately published more than twenty papers together. Percival Bailey, then on leave from the Department of Neurology and Neurosurgery at the University of Illinois, also contributed to the lab,[96] and he was to become instrumental in shaping the next phase of McCulloch's career. In parallel with his laboratory activities, however, McCulloch's scientific practices also involved an exploration of scientific philosophy.

## 3.8   Hull's Seminars and Scientific Philosophy

As numerous historians have shown, fluidity defined relations between the sciences and philosophy during the 1930s, particularly at Yale and Harvard.[97] Not only did philosophers interested in theories of knowledge look to ideal sciences such as physics and mathematics as a way of making philosophy more scientific, scientists also engaged with philosophical issues in order to secure a level of objectivity and rigor in both the content of and methods in their respective fields. Thankfully, historical assessments of the development of scientific philosophy treat logical empiricism not as an abstract, well-defined theoretical position but rather as a complex set of principles that must be viewed through the lens of institutional history.[98]

Thus, to attempt to present logical empiricism as a unified, coherent philosophical project would be misguided. However, many would agree that logical empiricists, in addition to valuing knowledge acquired through sense experience, aimed to use the tools of modern mathematical logic to account for the process and content of science.[99] Logical empiricism, by many accounts, was also a foundationalist project, as it involved for some a search for basic beliefs that would provide a foundation for a theory of knowledge or a scientific discipline. Overall, scientific philosophers aimed to recast the processes by which we acquire knowledge of the natural world, and historians of scientific philosophy highlight the rejection of traditional epistemology and an unraveling of Kant's synthetic a priori during this period, the latter of which began to be relegated to the realm of metaphysics.[100] The point of logical empiricism was not to improve science per se, but instill rigor and clarity to philosophy itself.[101]

In turn, not necessarily as a consequence, scientists imported methods of logic and a philosophical preoccupation with foundations to achieve similar aims for science and scientific method. Indeed, debates about "scientificity" were widespread across disciplines in the social, natural, and human sciences during this period. One science in which this convergence was most prominent was psychology. Laurence D. Smith has argued that the relations between behaviorist psychology and logical positivism can indeed be considered an alliance, yet the connection was not actually as close as traditionally argued. Rather, what characterized their relations were, on Smith's reading, "superficial convergences of opinion on broad issues and matters of rhetoric and propaganda."[102] While figures like neo-behaviorist Clark L. Hull, who was a prominent member of the Institute of Human Relations at Yale, indeed made contact with logical positivists in the late 1930s, he had already independently developed a scientific philosophy that was in line with their main ideals, particularly an emphasis on materialism, a rejection of metaphysics, deductive theorizing, and a commitment to the unity of science.[103]

As a psychologist, Hull extended John B. Watson's work, wherein all behavior is built up from reflex connections characterized as relations between stimulus and response. However, during the early 1930s, he became a proponent of the formal postulate method for developing psychological theory,[104] which he eventually outlined in both *Mathematical-Deductive Theory of Rote Learning* (1940) and *Principles of Behavior* (1943). In essence, Hull was advocating the hypothetico-deductive method of doing science, where a tentative hypothesis is adopted, its logical observable implications deduced, and observations then test the validity of the hypothesis.[105] As for the worth of this approach, while logical-mathematical principles are "merely invented rules of symbolic manipulation," logic is still worthwhile tool, for Hull, to derive "dependable expectations" in empirical investigations of the natural world:

The rules of logic are more dependable, and consequently less subject to question, presumably because they have survived a much longer and more exacting period of trial than is the case with most scientific postulates ... they come to each individual investigator ready-made and usually without any appended history. ... As a kind of empirical confirmation of the above view of as to the nature of logical principles, it may be noted that both mathematicians and logicians are at the present time busily inventing, modifying, and generally perfecting the rules of their disciplines.[106]

Appended to this last observation was a reference to Bertrand Russell's and Alfred North Whitehead's *Principia Mathematica*—just as they had aimed to formalize the rules of mathematics, Hull sought to do the same for psychology. Hull's goal was to develop basic principles from which empirical generalizations about behavior could be deduced, a foundation for a theory of behavior and learning.[107] This would achieve a level of objectivity for psychological science.

Hull's interest in machines reflected his mechanistic conception of behavior, wherein purposive behavior could be explained via mechanistic and automatic relations; there was no room for consciousness, free will, or any directive immaterial "mind."[108] Hull had a fascination with constructing "thinking machines" as a way of modeling and mechanizing the functions of intelligence and adaptive behavior, and many historians rightly point to this facet of Hull's science as an anticipator of the cybernetic project.[109] Yet as we shall see, Hull's importance for making sense of McCulloch's vision of cybernetics was less about thinking machines (although this is important) and more about foundations, a material basis for mind, and building scientific theory.

Unlike Watsonian behaviorists, Hull did not reject consciousness out of hand but treated it as something in need of explanation.[110] In his 1937 paper "Mind, Mechanism, and Adaptive Behavior," Hull argued that the problem of mind and adaptive behavior is not the subject of philosophical speculation but rather can be tackled using scientific methods—namely via the construction of a sound scientific theoretical system. Such a system should be full of clear postulates from which logically deduced theorems will arise. Such theorems should agree closely with observational facts of the scientific discipline.[111] A scientific theory with these characteristics would differ from a philosophical system in that it would be subject to observational test. Through such a system, Hull argued that consciousness could be integrated into a scientifically respectable theory of behavior.

In 1936, Hull began running a seminar on Wednesday evenings that was attended by diverse members of the Yale community. Devoted to hammering out an integrated research perspective for the IHR,[112] the series reflected Hull's main interest in scientific methodology and its relation to understanding behavior.[113] Seminar discussions were to be reflexive in that Hull aimed to eliminate both subjectivity and metaphysics in scientific practice.[114] Hull's vision also reflected a very hierarchical view of the relations between the sciences, with physics, logic, and mathematics forming the basis of a unified science. Hull stressed the importance and value of scientific theory and theorizing as a way of achieving disciplinary unity. His 1943 book *Principles of Behavior* was presented as part of a wider tradition in neo-behaviorist psychology to achieve a "systematic theoretical framework" for the study of behavior.[115]

Discussions at the seminars centered on interpreting psychoanalysis and behavior via deductive methods informed by logical empiricists. This attempt at rationalization was twofold: to come up with an objective *theory* of human behavior and a scientific *method* for the *study* of human behavior.[116] Hull characterized psychoanalysis as not only a therapeutic method but also as a conceptual system and body of facts.[117] Early on, the group explored psychoanalytic theory from a systematic standpoint in order to see if this framework was compatible with Hull's stimulus-response theory. Hull's aim was to determine whether psychoanalytic theory could be deduced logically from principles of behavior and learning theory.[118] Some members of this group attempted to formalize psychoanalytic theory in terms of precise definitions, postulates, and theorems. Seminar participants began exploring the libido in psychoanalytic theory and its "logical status" within that theory.[119] Later seminars continued to address assessing other theoretical approaches in psychology and the extent to which they had a formal structure.[120]

Among the participants at the seminars were Filmer S. C. Northrop and McCulloch. McCulloch began attending Hull's seminars in February 1936, and in March of that year presented papers on a comparison of phenomena of behavior in animals and human beings from a psychiatric perspective as well as regression.[121] According to Northrop's recollections, the seminar was devoted to "the theories and methods of mathematical physics" and it was here that McCulloch encountered Northrop's perspective on integrating neurological and psychological data into a meaningful theory of knowledge.[122] Frederic B. Fitch, at the time a doctoral student in the department of philosophy at Yale, discussed Russell's and Whitehead's *Principia* at Hull's seminars. Fitch was being taught by Northrop, and shared his view that logic had empirical content.[123] His view of logic was naturalistic and in line with that eventually developed by Hull in his elaboration of psychological theories.[124]

What did McCulloch take away from these seminars? McCulloch arguably adopted a perspective wherein his scientific and epistemological projects could merge. McCulloch was not oscillating between the laboratory and the seminar room, but rather the two sets of activities in each venue formed a unified project. Overall, his experimental work with Dusser de Barenne doing strychnine neuronography—mapping the functional organization of the cerebral cortex—taken together with his discussions on scientific foundations and knowledge, helped McCulloch develop an experimental epistemology: a physiological theory of knowledge. In the shadows of both Dusser de Barenne and Hull, the synthetic a priori (à la Magnus) could now be given scientific rigor, and not be thrown out as a metaphysical relic. While McCulloch's interest in the logic of the nervous system would not become explicit until the early 1940s, Northrop reports that this interest was fuelled by the atmosphere at these seminars.

## 3.9 Conclusions

In his 1937 Huxley Lecture, Nobel Prize–winning physiologist Edgar D. Adrian began by noting and celebrating the "successful invasion" of physiology into medical research and teaching.[125] Yet he lamented the fact that there were still many unsolved puzzles in linking the brain and mind. No detailed basis for memories and intelligence existed, nor were the roles of the cortex, thalamus, and hypothalamus in consciousness or emotional states clear. There was more to intelligent behavior, in Adrian's view, than the cerebral cortex. This was no admission of failure but a statement of the problems that faced cerebral physiology, and an emphasis on the fact that just such problems had fallen within the domain of neurophysiology. One of the successes of the 1930s on both sides of the Atlantic was that brain research was positioned as the basic component of clinical studies of the mind. Adrian was optimistic that the new methods of attack in understanding cerebral and mental organization were of great promise.

Adrian's reflections on the state of brain research notwithstanding, most American neurophysiologists during the 1930s pushed philosophical questions to one side. Yet as Susan Leigh Star has shown, brain researchers inevitably have certain philosophical assumptions about the brain-mind relationship, and such assumptions will inform the questions they ask about the brain and the methods they use to answer them.[126] To the extent that practices of cerebral localization were informed by philosophical assumptions, these were less about the "mind-brain" problem and more about particular models of brain function. The questions they posed about the brain were specific and experimentally tractable. Of course, they did ask questions about overall brain function and organization, but to the extent that they stepped back and offered a theoretical perspective, they developed theories of brain function that stressed functional organization, grounded in experimental study of the brain. Despite all the advancements in instruments and tools, which made study of the brain more precise than it ever had been, the brain was still an organ of staggering complexity, which humbled investigators dedicated to deciphering its secrets.

In the face of complexity, McCulloch sought objectivity, rationalization, and a firm foundation on which to build up studies of the mind and brain. During the 1930s at Yale, McCulloch performed the identity of a neurophysiologist, in a context where studies of the brain and the practice of cerebral localization brought a level of scientific objectivity to the understanding of human behavior. Here, McCulloch probed the brain using the material tools of the laboratory and the theoretical tools of integration and functional organization.

McCulloch's identity during this period was not an innate given nor was it simply a product of his institutional milieux at Columbia or Yale. Rather, through his dual practices and activities of laboratory work and philosophical discussion, McCulloch began to approach the development of a physiological theory of knowledge, an experimental epistemology. This was indeed transdisciplinary work, where philosophers were using scientific methods and scientists pursuing philosophical questions, questions of method and of knowledge. McCulloch performed his own idiosyncratic identity—that of an experimental epistemologist—through repeated laboratory and philosophical activities.

Dusser de Barenne died suddenly on June 9, 1940, of a coronary seizure while visiting Boston, and his death brought about a good deal of professional and financial uncertainty for McCulloch. In 1941, the fate of the Laboratory of Neurophysiology was unclear: should it be absorbed by the Department of Psychiatry or the Department of Physiology? McCulloch was asked his preference by Acting Dean of Yale's medical school, Francis Gilman Blake, but he did not seem to care either way, as long as he had a place to work and funds for the lab. As he wrote to Fulton, he was "going into the red to the tune of better than one thousand dollars a year" and now that Dusser de Barenne was gone, he knew he would need to look elsewhere for support.[127] He reported to Fulton that he'd been finishing work that he and Dusser de Barenne had been doing on the chimpanzee cortex, and had prepared material for "atlases" that was due to go to von Bonin in Chicago.[128] The budget for the lab was small, and McCulloch was obviously concerned about its future when he wrote to Percival Bailey in February 1941, wondering about what had been going on with respect to Bailey's "pipe dream" at the University of Illinois's Neuropsychiatric Institute.[129] Professionally, McCulloch was beginning to search for a new opportunity to continue his brain research. Clearly McCulloch thought Bailey's lab would be a good choice; and indicated so by sending along his CV and his rationale for doing neurophysiology:

[F]rom the time I was studying psychology and philosophy in college through the time I was working for my M.A. in psychology in [sic] Columbia—actually largely in psycho-pathology including work on Ward's Island—and all down these years in the lab my interest in psychiatry has been every present, for I am convinced we will get nowhere with crazy people until we can understand brains in such physico-chemical terms as we use when thinking of kidneys.[130]

Clearly at a crossroads in terms of his career path and in crisis mode, McCulloch evoked a parallel rationale for doing neurophysiology: psychiatry. Aligning his work on the brain and foundations with the growing field of American psychiatry was a choice that would serve him well. McCulloch hoped that any recommendation letters from Bailey could emphasize this point, and secure the "only kind of job in psychiatry that would ever interest me."

By spring 1941, McCulloch was successful, despite Fulton begging McCulloch to stay.[131] He joined the Illinois Neuropsychiatric Institute

(INI) in Chicago that summer, and became Professor of Psychiatry at the University of Illinois. The INI, having nine floors and a basement, was described as "the largest and most complete neurophysiological unit in the world."[132] As we shall see, here McCulloch's philosophical and physiological interests merged explicitly, in his bold confrontation of traditional neurophysiology—particularly its caution and pessimism when it came to drawing philosophical conclusions about the brain and mind.

# 4 The Egalitarian Mentor

## 4.1 Introduction

By the time Warren McCulloch had arrived at the University of Illinois in the summer of 1941, he had a decade of experimental work in neurophysiology under his belt, and regular yet marginal contact with like-minded scientists and philosophers interested in scientific methodology, logic, and the foundations of the sciences. The eleven years he subsequently spent in Chicago, which will be the subject of this and the next chapter, were the most transformative of his career. During the 1930s, neurophysiology had been defined by the need of scientific patrons—such as the Rockefeller Foundation—to address problems of mental illness with basic research on the brain, and in the early 1940s, this trend continued. When McCulloch arrived at the Illinois Neuropsychiatric Institute (INI), American psychiatry, still an eclectic field, was evolving even more closely with brain sciences; and at Illinois, McCulloch pursued work on the functional organization of the cerebral cortex in a more overtly clinical context.

Equally important were McCulloch's relations with younger scientists in the Chicago community. Now in his early forties, with more professional stability and security, McCulloch began to enter into personal, mentoring, and collaborative relationships that were to define his career for the remainder of his life, and to bring new scientific practices based on theoretical modeling to the forefront of his scientific work. No longer a research fellow, but a professor of psychiatry and eventually head of an entire research division, McCulloch had room to be bolder in his scientific aspirations. His generous, inclusive spirit meant that the junior scientists he collaborated with were less students than colleagues.

McCulloch engaged with two sets of scientific practices during this period, shaped by two facets of his institutional milieu. First, discussions with young mathematician Walter Pitts and Nicolas Rashevsky's Committee on Mathematical Biology at the University of Chicago, and to a lesser degree his association with Jerry Lettvin, reinforced McCulloch's early interest in theories of knowledge. Despite their differences in age and background, both McCulloch and Pitts were motivated by concerns that transcended the divides between science and philosophy. Through his association with a community of mathematical and theoretical biologists, McCulloch began to more overtly pursue scientific knowledge about the brain from a more formal, theoretical perspective. Like Hull and the logical empiricists, McCulloch and this group valued the scientific methods and perspectives of the physical sciences and aimed to generate foundational knowledge of living systems, in particular the nervous system.

Second, on a professional level, McCulloch continued his laboratory work on the functional organization of the cerebral cortex and addressed more clinical questions than he ever had at Yale. He became a neuropsychiatrist—a process examined more fully in chapter 5. While he adopted this persona for professional reasons, it began to shape the way McCulloch rhetorically presented his work on the brain.

Seemingly, then, McCulloch was pulled in opposite directions—between the messy, subjective, clinically motivated world of the laboratory and the abstract, pencil-and-paper world of mathematical biology. In order to gain a sense of how these two worlds intersected for McCulloch, here I focus on the institutional and intellectual contexts of his collaboration with Pitts that culminated in their 1943 publication "A Logical Calculus of Ideas Immanent in Nervous Activity," where they presented a theoretical account of the logical relations between idealized neurons, with purported implications for how the central nervous system functioned as a whole.[1] Traditional historical and philosophical examinations of this paper highlight its role in the history of artificial intelligence and the cybernetics movement, and McCulloch's activities during this period are often cast as a "pre-history" of his work as a cybernetician.[2] What is fundamentally more interesting about this work is how it emerged from a particular scientific context where science, philosophy, and medicine were being configured and brought together in new ways. My aim here is twofold. I first want to highlight the broad institutional and disciplinary context within which this collaboration took place. Second, I want to examine the ways that McCulloch's dual practices became intertwined in his work with Pitts. Their collaborative work emerged before anything resembling modern artificial intelligence existed, and before the cybernetics movement was well underway. While McCulloch's work during this period integrated philosophical themes, he was still engaged with more clinically motivated work and making contributions to studies of the functional organization of the cerebral cortex.

In this chapter, I set McCulloch's practices as part of a broad scientific project that saw clinically driven work on the brain and theoretical work on the brain as part of a more unified endeavor. By placing this story against a backdrop of mathematical and theoretical practices in biology and medicine, and the clinical context of the INI, I show that ultimately these practices, while at first glance at odds with each other, converged in McCulloch's rhetorical move: by going theoretical, and looking for foundational knowledge of the nervous system, one could purportedly present a scientific account of the mind (both healthy and diseased) and bring scientific rigor and objectivity to the discipline of psychiatry.

## 4.2   A Biological Basis for Psychiatry: The Illinois Neuropsychiatric Institute

Efforts to unite laboratory and clinical approaches to the brain continued to flourish in American psychiatry during the 1940s, as novel therapeutic innovations began to shape psychiatric practice. However, tensions existed between proponents of a more clinical approach to mental disorders and those that favored a more biomedical emphasis, as well difficulties establishing a disciplinary niche between psychiatry and neurosurgery.[3] Rockefeller Foundation support of psychiatry at the University of Illinois began as early as 1935, when Dean of Medicine D. J. Davis, on behalf of Harold Douglas Singer, approached Alan Gregg about funding for both clinical and basic teaching and research at Illinois.[4] Born in London, Singer had arrived in the United States in 1904 and in 1907 became head of the state-run Psychopathic Institute of Kankakee, Illinois, and state alienist (a specialist in psychiatry who treated "alien" states of mind). In 1919, he became professor of psychiatry at the University of Illinois.[5] Davis's proposal to the Rockefeller Foundation was grounded in Singer's views on psychiatry, and

Singer's 1935 proposal for undergraduate teaching in psychiatry at Illinois reveals much about his vision.[6] Singer viewed mental illness in the Meyerian framework, which related instances of maladjustment in patients to their living conditions, and thus when teaching psychiatry to undergraduates, he argued, emphasis must be placed on the whole person in relation to their environment. Overall, however, Singer saw a schism between the biological and dynamic approaches. More integration could be achieved not only by connecting basic instruction in disciplines that are fundamental to psychiatry (anatomy, physiology, and pathology) with clinical instruction, but also by incorporating the psychiatric approach in instruction in other fields such as surgery and obstetrics. Singer wanted to raise the profile of psychiatry within the medical school. His perspective aligned well with the Rockefeller program in psychobiology,[7] and in May 1936 the foundation granted the University of Illinois $45,000 over three years to develop teaching and research in psychiatry.[8]

Although the psychiatry of everyday life came to prominence during the early twentieth century, and became the basis of its expanding cultural power,[9] at Illinois, psychiatric research was directed by treatment of more serious mental and behavioral disorders.[10] By 1937, plans were well underway for a new neuropsychiatric institute, spurred on by Singer. The Illinois Neuropsychiatric Institute (INI) was to be jointly operated by the University of Illinois and the Illinois Department of Public Welfare, although the real driving force was the state. The Department of Public Welfare provided basic equipment for the institute, which was originally conceived as a state mental hospital having psychiatric and neurological divisions. The university was responsible for the INI's teaching and research, and the state funded the care of patients.[11] The INI was viewed as the "capstone" of the state mental hospital system. For the university, it was a prime opportunity for expansion of its program in psychiatry and especially research in neurology.[12]

The building for the INI, which cost $1.4 million (about $16.8 million in current dollars), was located at 912 South Wood Street, adjacent to the laboratories and teaching hospitals of the University of Illinois, on Chicago's West Side.[13] Rockefeller Foundation support of the INI, which began in November 1940, reinforced the close relationship between the medical school and the Department of Public Welfare.[14] Alan Gregg visited the new quarters for the INI in October 1940 with Percival Bailey, who had arrived at Illinois in 1939, and neurosurgeon Eric Oldberg, head of Neurology. The building itself consisted of two towers that held its two primary divisions: psychiatry and neurology/neurosurgery.[15] Gregg reported the "large division" for neurophysiology and experimental laboratories, with the neurological block having room for 54 patients, who were to be admitted from "any available source, including patients committed to State Hospitals under control of the Department."[16] The north tower housed surgical operating rooms, hospital floors, classrooms, and the offices of the Department of Neurology; the south tower housed the Department of Psychiatry. Research in the psychiatric division involved work using neurophysiological and biochemical methods. Ralph Gerard from the University of Chicago, who eventually joined the INI in 1952, was said to have designed and planned the neurophysiological labs, which were housed in the basement along with the neurological labs.[17] Having nine floors, the INI was said at the time to have been the largest neurophysiological unit in the world. While the INI included clinical facilities, its primary purpose was teaching and research. By 1942, the clinical branches of the INI had been reduced from four floors to two, having 38 adult patients and a children's ward on the sixth floor.

At the INI, the rhetoric surrounding the cultural and medical significance of psychiatry had shifted from Depression-era concerns about degeneration and self-control to gestures toward the human costs of World War II, which was especially prominent after the US was pulled into the conflict in December 1941. At the dedication of the institute on June 6, 1942, Illinois Governor Dwight H. Green observed,

We formally open this Institute at an hour in the world's history when it might be easy to despair of the fruits of science in human affairs. We have not found the way to conquer war, and we are seeing once again that war, the destroyer, either stifles scientific progress or uses science as a means of destruction. But when we turn our eyes to our American cities and states, there is still ground for hope in the ultimate triumph of science as a means of making life better for humanity.[18]

Scientific medicine now had the promise not simply to care for but cure the patient, and would reduce costs.[19] Casting the present conflict as a threat to democracy and a "violence of intolerance, hatred, and dictatorial lust," Governor Green presented the INI as having a huge stake in this "war for survival"—neuropsychiatry had the potential to save democratic science and address the problem of war head-on.

Tragically, Singer had died in August 1940 of a pulmonary embolism following a car accident, and finding his successor was not easy. But in 1941,
Francis T. Gerty became the new head of the Department of Psychiatry, and
like Singer, he believed in a close professional and intellectual resonance
between the problems of psychiatry and the fields of chemistry, physics,
physiology, psychology, and sociology.[20]

McCulloch's perspective on the relationship between neurophysiology
and psychiatry was in perfect alignment with Gerty's, and he joined the INI
at an exciting time for Illinois psychiatry.[21] In December 1941, he wrote to
Dusser de Barenne's daughter, bursting with enthusiasm, reporting that the
lab there was "the best designed I have ever seen. ...."[22] McCulloch endorsed
Singer's original vision for the INI, based on his belief that "we would get
nowhere with psychoses until we worked out their biological foundation,
which is, after all, the physiology of the central nervous system, biophysics
and biochemistry included."[23] By the early 1940s McCulloch was an expert
in the functional organization of the cerebral cortex,[24] and when he arrived
at the INI, he began collaborating with Bailey in Neurology and Gerhardt
von Bonin in Anatomy on the structural and functional divisions in the
cerebral cortex of the chimpanzee (particularly the occipital lobe, involved
in vision). They continued to use Dusser de Barenne's strychnine technique, and unlike the more stern, intense atmosphere at Yale's Laboratory
of Neurophysiology, the atmosphere in the research laboratory at the INI
was informal and lively.[25]

McCulloch's disciplinary framework for making sense of the brain had
clearly shifted since his days at Yale. At the INI, researchers pursued basic
research on the brain with more explicit psychiatric motivations and this
work was carried out *within* a Department of Psychiatry. The clinical
research on the causes, diagnosis, and treatment of mental disease strongly
correlated with "basic investigations."[26] In a prospectus penned circa 1942,
Gerty outlined the principles that were to direct the organization of and
practice in the research laboratory for the next ten years.[27] For Gerty, work
in the laboratory would provide a biological foundation for a scientific
approach to disorders of the nervous system. The best approach to laying
such a foundation was interdisciplinary "team play" that brought together
workers in physiological psychology, neurophysiology, anatomy, physics,
and biochemistry. Gerty believed the work in the lab should be in "intimate
contact with the needs of patients," and thus saw the need for three new

posts: a seasoned psychiatrist, an electroencephalographer, and a psychologist—who could act as intermediaries between the laboratory and the patients. The INI's neurophysiology laboratory had McCulloch as a neurophysiologist, J. R. Klein as biochemist, Craig W. Goodwin as a physicist, several technicians and administrative personnel, and several students, including medical student Jerome Y. Lettvin.

## 4.3  McCulloch, Pitts, and Lettvin

Although the McCulloch family moved to Hinsdale, Illinois, a rural area on the outskirts of Chicago, they continued to maintain and work the farm in Old Lyme. In this rural atmosphere, McCulloch and Rook raised their children to be self-reliant and responsible—chores around the house each morning before school, and as soon as they were old enough, the children were farm hands—pumping water, feeding livestock, harvesting corn. The children all called their parents "Warren" and "Rook." The open, collective spirit that had originated at Camp Aladdin continued at Old Lyme and was transferred to their home in Hinsdale. McCulloch was an inclusive man; never hiding behind credentials, he spoke to young scientists not as children but as experts in whatever they happened to be discussing, and listened "as carefully to a child as to a colleague."[28] This openness and egalitarian approach to mentoring was an important facet of McCulloch's legacy, and, I argue, was crucial to McCulloch's practices at Illinois outside the laboratory. Here he began to enter into scientific and personal relationships with two men who were to become his most important collaborators: Jerome Lettvin, who was twenty-one when he met McCulloch, and Walter Pitts, who was sixteen. Lettvin had had entered medical school at Illinois in the fall of 1939.[29] He graduated with a B.S. in 1942 and an M.D. in 1943, and by his third year of medical school was introduced to the INI and McCulloch by von Bonin. Lettvin recalled his first meeting with McCulloch:

When I was in high school, I had read Buckle's history of the Scotch mind in the seventeenth century. It gave me nightmares for weeks afterward. Warren would be the first Scotchman I ever knew, and truly, I had a frisson going to see him. This was not lessened when I met him. He sported no beard then, but certainly had the most piercing eyes I ever encountered. However, in very little time, he, acting most friendly, dispelled my nervousness.[30]

Ultimately, Lettvin joined the INI and began to work with McCulloch on the study of the nervous system. The previous year, Lettvin had met Walter Pitts [figure 4.1], who was to become his lifelong friend. Pitts was fifteen at the time.

Lonely, impoverished, and brilliant, Pitts's tragic story, largely based on the recollections of Lettvin, has been told many times.[31] A polymath and autodidact who had taught himself Latin, Greek, and Sanskrit, Pitts had an early interest in mathematical logic. Lettvin recalled Pitts's encounter with the work of Bertrand Russell:

At the age of twelve [Pitts] was chased into a library by a gang of ruffians, and took refuge there in the back stacks. When the library closed, he didn't leave. He had found Russell and Whitehead's *Principia Mathematica*. He spent the next three days in that library, reading the *Principia,* at the end of which time, he sent a letter to Bertrand Russell, pointing out some problems with the first half of the first volume; he felt they were serious. ... A letter returned from Russell, inviting him to come as a student to England—a very appreciative letter. That decides him; he's going to be a logician, a mathematician.[32]

Pitts apparently met Russell in 1938, when he was on sabbatical at the University of Chicago. Through Russell, Pitts met scientific philosopher Rudolf Carnap, who was in the Department of Philosophy. Pitts began to attend classes at Chicago, and soon began a close association with Nicolas Rashevsky's mathematical biology group at Chicago.

## 4.4   Rashevsky and Mathematical Biophysics

Born in Chernigov, Ukraine, in 1899, Nicolas Rashevsky was trained in theoretical physics at the University of Kiev.[33] Immigrating to the United States in 1924, he worked as a physicist at the Westinghouse Research Laboratories in Pittsburgh, and as a lecturer in physics at the University of Pittsburgh. Rashevsky had published several papers on a mathematical theory of nerve conduction during the early 1930s, built on the notion of diffusing substances and electrochemical gradients.[34] This was a common approach in mathematical treatments of conduction and transmission, where theorists used differential equations to describe the relations between intensity of stimulus and concentration of ions.[35] In 1934, Rashevsky was invited to the University of Chicago as a Rockefeller Fellow, a move facilitated by several members of the Chicago community, including

**Figure 4.1**
Walter Pitts, circa 1950. McCulloch Papers, Folder "Walter Pitts," Box 4, Series VIII
Photographs, WSM Papers, APS.

physiologist Ralph S. Lillie and neuropsychologist Karl Lashley. In light of Chicago's institutional prominence in biology, biophysics, and neurophysiology, Rashevsky's association with Chicago is not surprising.[36] By 1935, Rashevsky joined the Department of Physiology.

Unlike others of his time, Rashevsky was not simply attempting to use theoretical and mathematical methods in a discipline that for the most part involved experimentation, but aimed to develop a scientific field devoted to such pursuits. His brand of mathematical biology was analogous to mathematical physics, related to experimental biology in the same way in which mathematical physics stood to experimental physics. Mathematical treatments in biology did exist at the time: in the preface to his *Mathematical Biophysics* (1938), Rashevsky cited D'Arcy Thompson's *On Growth and Form* (1917), as well as the work of Alfred J. Lotka[37] and Vito Volterra[38] on species interaction in a population of organisms.[39] However, in contrast to work in physiology and neurophysiology during this period, which was highly empirical, Rashevsky's approach to biological problems was almost entirely theoretical, full of idealizations and mathematical equations, and had little contact with experimental work. Rashevsky drew on what he viewed as the fundamental practices that defined mathematical physics, where complex phenomena were simplified and idealized mathematically:

Following the fundamental method of physicomathematical sciences, we do not attempt a mathematical description of a concrete cell, in all its complexity. We start with a study of highly idealized system. ... We ordinary mortals must be more modest and approach reality asymptotically, by gradual approximation.[40]

By the late 1930s, Rashevsky began an independent group for "mathematical biophysics" at Chicago and eventually, through the aid of the University of Chicago and the Rockefeller Foundation, he created a doctoral program in mathematical biology, and his group became known as the Committee on Mathematical Biology. By March 1939 Rashevsky founded a new journal, the *Bulletin of Mathematical Biophysics*,[41] which was a forum for mathematical treatments of psychological and neurological phenomena, and eventually served as a principal publication outlet for most mathematical biologists, and for much of Rashesvky's own work.[42] Through this journal, however, Rashevsky's publications became ghettoized. While reviewed by important publications in physiology and neurophysiology, on the whole the relevance of Rashevsky's work was lost on mainstream physiologists.[43]

In terms of discipline-building, Rashevsky did not find clear success. However, less formally, he influenced a variety of budding scientists and theoreticians. Toward the end of the 1930s, Rashevsky began holding weekly seminars on Friday afternoons that included participants from both the University of Chicago and the University of Illinois. They were affectionately referred to as "samovar" meetings because of the large antique Russian samovar that stood in the seminar room.[44] Around 1940, sixteen-year-old Walter Pitts began attending these seminars—where he would often point out errors in Rashevsky's reasoning, and Rashevsky ultimately took Pitts in as part of his group.[45] One list of invitees from circa 1941 included Ralph Gerard, Sewall Wright, Ralph Lillie, and Gerhardt von Bonin.[46]

Sometime in early 1941, Lettvin took Pitts to meet McCulloch[47] and, as Lettvin recalled, given McCulloch's philosophical interests, "[t]here was no question at all but that Walter instantly became part of Warren's coterie."[48] The relationship was to change both men. McCulloch also began attending Rashevsky's seminars in winter 1941, regularly making his way south from Wood Street to the University of Chicago campus near Hyde Park. In the context of these seminars, McCulloch, Pitts, and Lettvin evolved as a collaborative group. Early in 1942, Pitts and Lettvin moved into the McCulloch home.[49] Always ready to accept wayward and troubled youth into their home, the McCulloch generosity extended into his professional life. As Rook McCulloch later recalled, McCulloch had an "ease of communication" with people of every age.[50] Throughout the 1940s, this was a free-spirited environment, and the household was always crowded with houseguests coming and going, many from McCulloch's group at the University of Illinois. Barbara Wiener had stayed with the McCulloch's for some time during 1947, and recalled, with some derision: "The laird of the castle appeared and disappeared at will, talking grandly and holding court. The household was proudly radical. The phonograph in the living room kept spinning out songs of the Spanish Civil War and union songs to which everyone knew the words but me."[51]

Although McCulloch was more than twenty years Pitts's senior, and was the driving force behind the collaboration, their relationship was not a typical one of mentor to student; instead they were intellectual peers.[52] As I have argued elsewhere, McCulloch's style of mentoring both Pitts and Lett-vin was more "laissez-faire" than disciplinarian, and this colored the per-sonal and intellectual alliances he was to have with both men.[53] Pitts and Lettvin ultimately became McCulloch's most important scientific collabo-rators, and their relationships, defined by emotional and intellectual inten-sity, were to define the lives of all three men for the next thirty years. While ensconced in the world of neuropsychiatry, McCulloch's weekly escapes to Rashevsky's group and his excited conversations with Pitts and Lettvin pulled him into a new, and for now parallel, scientific world.

## 4.5   Mechanizing the Mind: The McCulloch–Pitts Collaboration

In May 1942, after being at the INI for less than a year, McCulloch attended a small meeting on "Cerebral Inhibition" at the Hotel Beekman on Park Avenue in New York City. Sponsored by the Macy Foundation and orga-nized by medical director Frank Fremont-Smith, the meeting brought together a diverse group of researchers by invitation only, including Har-vard physiologist Arturo Rosenblueth, social scientist Gregory Bateson, anthropologist Margaret Mead, and psychiatrist Lawrence Kubie.[54] The planned topics for the three-day event were the physiology of the conditioned reflex and hypnosis.[55] Experimental psychologist Howard S. Liddell was to speak on the conditioned reflex, with neurologist Harold G. Wolff offering up the use of his laboratory at the New York Hospital for discussions of hypnotic phenomena. All participants, save for Bateson and Mead, came from the fields of psychology, psychiatry, neurology, or neurophysiology.

While this meeting has been touted by many as the first formal encoun-ter of some of the figures who eventually became members of the cyber-netics group, what is more relevant here is what McCulloch's response to the meeting reveals about his scientific practices. Most striking for McCulloch—and for the other participants—was Mexican-born Harvard physiologist Arturo Rosenblueth's presentation, which was based on dis-cussions that he'd been having with MIT mathematician Norbert Wiener. Rosenblueth's talk was expanded into a 1943 publication with Wiener and engineer Julian Bigelow. Rosenblueth and Wiener had met while Rosen-blueth was a Visiting Researcher at Harvard and while Wiener had been doing government-sponsored research on the anti-aircraft predictor.[56]

Rosenblueth, Wiener, and Bigelow concluded that "all purposeful behavior may be considered to require negative feed-back." They also claimed that their classification not only emphasizes the concepts of purpose and teleology, but also shows them to be important. By aligning teleology with purposeful behavior, Rosenblueth, Wiener, and Bigelow argued they have presented a more scientifically respectable alternative to traditional discussions of teleology—causality and final causes are abandoned.[57] They also saw merit in their classification because it allowed for a uniform behavioristic analysis applicable to both organisms and machines, *regardless of the complexity of the behavior*. For Rosenblueth, Wiener, and Bigelow, the role of negative feedback was in a sense to "rescue" teleology and make it and the concept of purpose more respectable and legitimate in discussions of living systems. Further, in their view, their focus on negative feedback and on behavior provided a way of transcending the divide between organisms and machines—in any domain in which purposeful behavior occurs, processes of negative feedback are at work.

Rosenblueth's talk and the meeting itself generated a great deal of excitement among the participants, particularly McCulloch. A few weeks later, McCulloch wrote to Fremont-Smith with ideas for papers he'd been working on, and the ways that he'd been inspired by discussions at the meeting.[58] He first outlined the gist of a paper on the "formal" and "final" aspects of "mentality"—the formal properties of the nervous system and the relations between "neuronal reactions." McCulloch was strongly inspired by Rosenblueth's talk:

I am in entire agreement with Rosenblueth as to the dependence of "goal directed" behavior upon "feed-back" mechanisms. From them it can be deduced. I think much discussion as to the nature of these "feed-back" systems is in order. They exhibit such phenomena as unwanted circuit actions and the phenomenon of hunting which is sufficiently intense to lead to destruction of the device in question. I do not for a minute think that the physical understanding of any oscillating disturbance is a simple matter; but these are sufficiently familiar and *so undeniably physical that if we can convert the problem of neuroses or psychoses into terms of these types of mechanisms, we will be well on our way towards thinking about them sensibly—I mean mechanistically and physically* [my emphasis].[59]

At this stage, for McCulloch, the worth of conceiving of mental phenomena in terms of feedback was the potential for physical, mechanistic, even biological—and thus respectable accounts of the mind and its pathologies. Not only did the Rosenblueth-Wiener-Bigelow paper discuss a mechanical account of goal-directed phenomena, for McCulloch, such an account held promise for a formalization of mental activity (either normal or abnormal—McCulloch's gesture to "unwanted circuit actions") and a pursuit of the foundations of brain sciences, in the tradition of Clark Hull. Overall McCulloch very much hoped the group would meet again.[60]

According to his autobiographical reflections, McCulloch had already attempted to formulate a logical calculus to describe the all-or-none activity of neurons during his time at Yale in the 1930s, but he had no strong foundation in mathematical logic. Whether or not we can take McCulloch at his word, it is also likely that McCulloch's views on logic could not help but be shaped by the work of Hull, Northrop, and Fitch, who all believed logic has empirical content. Thus McCulloch sought to pursue the logic of relations naturalistically—the laws of logic are the laws of thought.[61] Pitts and McCulloch quickly discovered they had an intellectual connection in their interests in mathematical logic. According to Lettvin, Pitts had been reading the work of seventeenth-century German philosopher and mathematician Gottfried Leibniz, who had related the notions of computation, logic, and algorithms,[62] and had demonstrated that "any task which can be described completely and unambiguously in a finite number of words can be done by a logical machine."[63] Much later, McCulloch said that it was British mathematician Alan Turing's idea of a "logical machine" that had inspired them. In 1936, Turing had developed a theoretical machine for the process of mathematical computation.[64] Simply put, he defined the complicated process of computation in "mechanical" terms, with the notion of a simple algorithm so exhaustive and unambiguous that the executor would need no mathematical knowledge or expertise to carry out its task. In characterizing computation this way, Turing introduced "states of mind" that were irreducible: the operations performed by a logic machine or a human computer could be split up into "simple operations" so elementary they cannot be further divided. This concept was central to the McCulloch–Pitts collaboration, and the notion of a logical machine fascinated Pitts.

The relations between McCulloch, Pitts, and Lettvin were intense during this period. The three, particularly Pitts and Lettvin, were almost inseparable—Pitts divided his time between the McCulloch home in Hinsdale, Lettvin's room, and "his own shabby apartment in Hyde Park."[65] In appearance and temperament, Pitts and McCulloch were almost opposites: Pitts was slight, quiet, and introverted (yet adventurous); McCulloch was tall, intense, animated, and flamboyant. Although we know little of their process of writing their most famous paper, Lettvin recalled that most of the work was done "in the quiet of the night, after McCulloch's own children, who were not much younger than Pitts, had to be put to bed. ..."[66] The main gist of the paper was hammered out in a matter of weeks. By the end of 1942, McCulloch and Pitts had completed the essay that was to become their famous 1943 paper, with Pitts contributing most if not all of the technical skill in logic.

Beyond this element of mathematics and logic, McCulloch's work with Pitts brought in an interest in the logic of human thought. When McCulloch had purportedly asked, "What is a man, that he may know a number?" he was pondering the nature of mental activity and its physiological basis in the brain. As he recalled: "My object, as a psychologist, was to invent a kind of least psychic event, or 'psychon.'"[67] McCulloch was not the first to discuss the notion of a psychon. The concept had emerged during the first decades of the twentieth century in neurology and psychology.[68] Generally, in material terms, the psychon was a unit of matter related to mental functioning. It signified the atomism characteristic of the modern era, the disciplinary aspirations of psychology as a natural science, as well as a bridge over the gap between material and immaterial accounts of the mind.[69] This last aspect had the most relevance for McCulloch, and the concept was tied up with his interest in logic:

[the psychon] would have the following properties: First, it was to be so simple an event that it either happened or else it did not happen. Second, it was to happen only if its bound cause had happened ... that is, it was to imply its temporal antecedent. Third, it was to propose this to subsequent psychons. Fourth, these were to be compounded to produce the equivalents of more complicated propositions concerning their antecedents.[70]

McCulloch later explained that a "psychon," for him, was a "simplest psychic act"—"what an atom was to chemistry, or a gene to genetics. ... But my psychon differed from an atom and from a gene in that it was to be not an enduring, unsplittable object, but a least psychic event."[71] The notion of an event occurring "only if its bound cause had happened" and proposing this to "subsequent psychons" implied the notion of a network of logically connected elements governed by relational rules. Soon McCulloch realized that these could be conceived as the all-or-none impulses of neurons.[72]

The all-or-none principle, formulated by British neurophysiologist Edgar Douglas Adrian in 1912, came from laboratory work on the peripheral nervous system. The principle described an experimental phenomenon and provided a framework for making sense of the transmission of nerve impulses. In his first published results on the topic, Adrian had observed that when a nerve impulse went through the narcotized section of a nerve, the impulse progressively lost strength.[73] Adrian found that a nerve impulse could recuperate after emergence from an area of narcotization, even if the strength of the stimulus had slightly "petered out." The magnitude of the impulse just before recovery had no influence on its magnitude after recovery. Adrian concluded that the propagation of the impulse was independent of the strength of the stimulus, and that the relationship between stimulus and response is all or none.[74] In 1925, working with the Swedish physiologist Yngve Zotterman, Adrian obtained pictorial confirmation of the all-or-none relation between stimulus and response in single nerve fibers, demonstrating that each fiber responded *fully* to a stimulus if it responded at all.[75] Adrian's work on characterizing the nerve impulse earned him a Nobel Prize in 1932, shared with the eminent neurophysiologist Charles Scott Sherrington. As Robert Frank has observed, how nervous function was understood had been transformed from an abstract, verbal, qualitative principle to a set of images created by instruments;[76] one of the many sorts visual inscriptions that had been central to electrophysiology since the nineteenth century.

In McCulloch's work with Pitts, logic, the notion of a psychon, and the all-or-none principle were brought together in a theory that transcended the divides between science, philosophy, and experimental medicine. McCulloch and Pitts were not interested in the mechanism of transmission of the nerve impulse but rather in the *relation* between stimulus and response, which they ultimately translated into a *logical* relation. Generally speaking, logic is the theoretical study of the structure of reasoning, the analysis of the language of propositions, and their logical relations.[77] Logic became mathematized during the late nineteenth century, when the rules of language, reasoning, deduction, and inference—operations of the mind—were mathematized with the development of mathematical logic, an "algebra of logic." The logic of propositions can be symbolized, with variables or symbols representing propositions or sentences, resulting in what we now call Boolean algebra, after George Boole, the mathematician who contributed to its development.[78] Propositions in logic are simply sentences or statements that are free of any ambiguity, and, in certain contexts, are seen as either true or false but not both. Three typical functions (sometimes called Boolean functions) in propositional logic are conjunction (AND, symbolized by "·"), disjunction (OR, symbolized by "∨"), and negation (NOT, symbolized by "~"). In light of his discussions with Pitts on logic, McCulloch observed that as propositions in propositional logic can be "true" or "false," neurons can be "on" or "off"—they either fire or they do not. This formal equivalence led McCulloch and Pitts to argue that the relations among propositions can correspond to the relations among neurons, and that neural activity can be represented as a proposition.

By the end of the 1930s, relations between individual nerve cells and the transmission of the nerve impulse, once restricted to studies of the peripheral nervous system, began to take on a new relevance within the context of studies of the cortex. Neurophysiologists such as Rafael Lorente de Nó at the Central Institute for the Deaf in Washington University in St. Louis had proposed that there were "chains" of neurons in the central nervous system that formed closed circuits around which nerve impulses may circulate indefinitely.[79] Histological evidence had revealed and confirmed the presence of such chains,[80] and researchers began to explore the relationship between the cytoarchitectonic structure of the cortex—the pattern and distribution of cell types—and the electrical patterns recorded using the microelectrode and cathode ray oscilloscope. In the words of psychologist Edwin Boring, "it was supposed that [in the brain] the fibers merely formed a complicated network ... and that the physiological account of mind was somehow to be gained from a further knowledge of this network."[81]

The 1943 paper that resulted from McCulloch's and Pitts's first collaboration resembled nothing before it. While prominent Manhattan psychiatrist Lawrence Kubie had explored the possibility of explaining spontaneous neurological activity through hypothetical circular pathways along which nervous impulses move, and privately McCulloch cited his work as the "first" to propose reverberations in the central nervous system,[82] it was not presented as a formal theory nor was it grounded in logic.[83] Published in Rashevsky's *Bulletin of Mathematical Biophysics*, the McCulloch–Pitts paper sat at the boundaries between disciplines—simultaneously addressing questions in neurophysiology, logic, and psychiatry. Scientifically, it modeled neural activity and amounted to an argument about the logical possibilities of how neurons could be functionally organized. Yet these questions were pursued theoretically and formally, with conclusions that had implications for understanding the mind and its pathologies.

The structure of the McCulloch–Pitts paper is clearly in the vein of Hull's deductive methodology for grounding psychological theory. McCulloch and Pitts began their paper with generally accepted knowledge about the nervous system. First they assumed that the nervous system is a network of neurons, each having a soma and an axon, with synapses between the axon of one neuron and the soma of another. They also assumed that at any instant a neuron has some threshold, which excitation must exceed to initiate an impulse. Third, they assumed that excitation occurs mainly from an axon termination to a soma, and inhibition occurs when the activity of one neuron or group of neurons is prevented by the activity of a second neuron or group. Their final assumption was that excitation across synapses occurs mainly from axonal terminations to somata.[84] None of these assumptions departed from knowledge generated by empirical investigation in neurophysiology. However, McCulloch's and Pitts's goal was to move beyond empirical evidence and represent the functional relationships between neurons in terms of Boolean logic: to embody reasoning in the physiology of the brain. To do this, they needed to make certain theoretical presuppositions. Most notably, they presupposed that the activity of a neuron is an all-or-none process and that the structure of the net does not change with time.[85] While McCulloch and Pitts admitted that this was an abstraction, they emphasized that their goal was not to present a *factual* description of neurons, but rather to design "fictitious nets" composed of neurons whose connections and thresholds are unaltered.[86] [figure 4.2]

In part, their conception of the central nervous system drew on McCulloch's practices in cerebral localization. McCulloch's 1943 publication with Pitts embodied a key feature of the cybernetics movement: the application of mathematical, formal methods to problems in the biomedical sciences; institutionally speaking, this aligned well with the practices of the Rashevsky group. However, several elements came together in the McCulloch–Pitts paper that undermine this simple reading. First, the work brought work on mathematical logic and problems in computability to

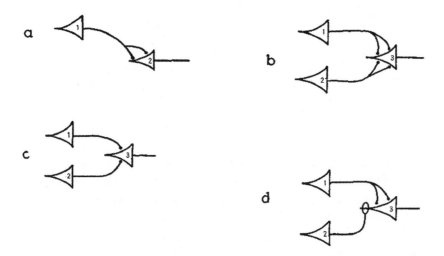

**Figure 4.2**
Examples of the McCulloch–Pitts neural arrangements and their corresponding expressions in propositional logic. In each example, a sum of two excitatory synaptic connections (represented in the diagram by dots adjacent to neurons) is required for a neuron to fire. An inhibitory connection is represented by an open circle adjacent to the neuron. In Figure (a), neuron 2 will fire if and only if neuron 1 fires. Logically, this corresponds to the expression $N_2(t) \equiv N_1(t\text{-}1)$, which can be read as "neuron 2 will fire at time ($t$) if and only if neuron 1 fires at time ($t\text{-}1$)." Figure (b) shows a network that is isomorphic with the Boolean function "OR" in propositional logic. Its expression, $N_3(t) \equiv N_1(t\text{-}1) \vee N_2(t\text{-}1)$ means that neuron 3 will fire at time ($t$) if and only if neuron 1 fires $or$ neuron 2 fires at time ($t\text{-}1$). Figure (c) demonstrates the Boolean "AND" function. The expression $N_3(t) \equiv N_1(t\text{-}1)\ N_2(t\text{-}1)$ means that neuron 3 will fire at time ($t$) if and only if neuron 1 fires at time ($t\text{-}1$) $and$ neuron 2 fires at time ($t\text{-}1$). McCulloch and Pitts also provided an example of the Boolean "NOT" function, with the instance of an inhibitory neuron. In the logical expression corresponding to Figure (d), $N_3(t) \equiv N_1(t\text{-}1) \sim N_2(t\text{-}1)$, means that neuron 3 will fire at time ($t$) only if neuron 1 fires at time ($t\text{-}1$) and neuron 2 $does\ not$ fire at time ($t\text{-}1$). From Warren S. McCulloch and Walter Pitts, "A logical calculus," 130.

bear on our understanding of the brain.[87] Second, it incorporated modern notions of the mind where thought was rendered atomistically and mechanistically. Building on, yet departing from McCulloch's extensive training in experimental neurophysiology, this work began with formal assumptions, theoretical presuppositions, and idealizations, based on general knowledge of the nervous system and nerve cells. McCulloch and Pitts translated the all-or-none principle into the supposition that just as propositions in propositional logic can be "true" or "false," neural cells can be "on" or "off"—they either fire or they do not. The "all-or-none" law, in their words, ensured that "the activity of any neuron may be represented as a proposition."[88] What implications did McCulloch and Pitts see for this picture of neural activity?

## 4.6   The Mind No Longer Goes "More Ghostly than a Ghost"

Although the only citations in the paper are to mathematicians and logicians, in the first and last sections of the paper, McCulloch and Pitts discuss the implications of their model for physiology and psychiatry. They presented a picture of brain activity that was deterministic and to some extent predictive—as long as the network presented was unaltered. They concluded that both the logical and purposive aspects of mental activity are "deduceable" from neurophysiology:

The psychiatrist may take comfort from the obvious conclusion concerning causality—that, for prognosis, history is never necessary. He can take little from the equally valid conclusion that his observables are explicable only in terms of nervous activities, which, until recently, have been beyond his ken. The crux of this ignorance is that inference from any sample of overt behavior to nervous nets is not unique, whereas, of imaginable nets, only one in fact exists, and may, at any moment, exhibit some unpredictable activity. Certainly for the psychiatrist it is more to the point that in such systems "Mind" no longer "goes more ghostly than a ghost." Instead, diseased mentality can be understood without loss of scope or rigor, in the scientific terms of neurophysiology.[89]

There are several things going on here. Clearly, McCulloch's work with Pitts was prompted by a firm belief in somatic psychiatry and impatience with a more psychodynamic approach to treatment. Secondly, McCulloch wished to bring objectivity to psychiatry as a discipline. To make the discipline scientific, one must develop an objective theory of mind, achieved only through theoretical foundations drawn from knowledge in neurophysiology. There were disciplinary stakes in discussions of the mind-brain question—with McCulloch's ease of moving across disciplinary boundaries, he was expressing frustration at their existence.

When viewed in light of the recent Cerebral Inhibition meeting, McCulloch also aimed to bring scientific respectability to discussions of purpose in living systems by mechanizing seemingly goal-directed phenomena. Traditionally, teleological or purposeful phenomena had served as an argument for the autonomy of the life sciences from the physical sciences. Further, certain apparently goal-directed phenomena were often explained by referring to some immaterial organizing principle, such as Hans Driesch's entelechy for embryonic development. For McCulloch, with his disdain or psychoanalysis (and perhaps terms like "the unconscious"), one must eliminate the idea of an immaterial mind.

McCulloch's and Pitts's reference to the notion of mind going "more ghostly than a ghost" was a gesture to the views of Charles Scott Sherrington. As we have seen, Sherrington was a towering figure in Anglo-American neurophysiology, and his influence on the conceptual foundations and laboratory practices of American neurophysiology was unequalled. In 1937–1938, he delivered the prestigious Gifford Lectures in Edinburgh, which were published as a book in 1940 entitled *Man on His Nature*. The book is a work of broad historical, scientific, religious, and philosophical scope, and aimed to present an account of the past and present sciences within a humanist framework. Sherrington's perspective was typical of interwar Britain, where tension existed between those embracing the promise of progress based on science and technology and those who pushed for a rebirth of a lost Victorian consciousness, where moral and spiritual goals took precedence over material and scientific ones.[90] Here Sherrington presented a fairly straightforward dualist position on the question of mind and its relationship to the brain.[91]

Roger Smith has offered a close analysis of Sherrington's brand of "dualism" and debates in the British context about the relationship between mind and body, where "mind" stood as a cultural as well as scientific term.[92] On Smith's reading, these mind-brain discussions were actually subordinate to debates about cultural values, with dualism aligning with a conservative position. Smith also points to a tension between Sherrington's dualism and his hope for an "integrated mind-body science."[93] Since Sherrington believed that it was impossible to say anything about how the mind *interacts* with the brain, he can be deemed a dualist. However, according to Smith, in research in the 1930s and 1940s on synaptic conduction, the EEG, and the "mid-brain functions," the notion of integration was everywhere (translated into discussions of "pattern" and "organization") and many brain investigators were working in the integrative framework that Sherrington had outlined in 1906 in his influential book *The Integrative Action of the Nervous System*. Smith explains this tension by arguing that

Sherrington's dualism was one of "cultural life rather than ontology." Sherrington's ideas about mind came from a conservative British culture that drew upon resources from religion, philosophy, the arts, and the classics.[94] Mind, for Sherrington, was not a religious soul, but rather as part of the "individuality" of the organism: it represented the values of a humanistic culture. This was what Sherrington was defending with his dualism, not an entity "outside the scope of science." While his notion of integration reflected his hopes for integration between psychology and biology, his dualism was connected to his respect for disciplinary divisions between the sciences and the humanities.[95]

McCulloch's attitudes about the brain and mind also reflected his own perspective on the relations between science, medicine, and philosophy. Each discipline, for McCulloch, presented a set of tools that could be used interchangeably and at times simultaneously. Seen within this context, McCulloch's anti-dualist position on the mind-brain question also reflected his values and disciplinary perspective. McCulloch believed that Sherrington had denied the possibility of a biological basis of mind,[96] and in a sense McCulloch's criticism of Sherrington came from a distinct perspective on disciplinary relations: Sherrington believed natural science should be "silent" on consciousness, while McCulloch wished to transcend the divide between the natural and human sciences.

The problem of mind and brain had always been an empirical, scientific question for McCulloch. In May 1943, he had presented a discussion at the Illinois Psychiatric Society, following a paper given by psychoanalyst Franz Alexander on psychosomatic research.[97] Alexander played a fundamental role in developing the field of psychosomatic research in Chicago. Besides being founder and head of the Chicago Institute for Psychoanalysis, he was also a faculty member in the Department of Psychiatry at Illinois. Alexander had remarked that he hoped to do away with "all discussion of mind-body problems from the publications of psychosomatic medicine." McCulloch was in total agreement—not because he was not interested in the problem, but because he was frustrated with the traditional mind-body dichotomy itself. He pointed to what he saw as a dichotomy in medicine, with a psychiatric approach to one side, with the use of "mentalistic" terminology, and the "organic" approach on the other, with the use of physical terminology. McCulloch lamented that "it remains our great difficulty that we have not ever managed to conceive how our patient ... can have a psychological aspect and a physiological aspect so divorced."[98] In his discussion, McCulloch pointed to two recently published books that exemplified the historical failure to unite mental processes with physical ones: the first was Sherrington's *Man on His Nature* (1940), and the second was the 1938 book *The Place of Value in a World of Fact*, by the Gestalt psychologist Wolfgang Köhler, which can be placed within discussions and debates on the topic of "science and values" that dominated American intellectual life in the 1930s.[99] Gestalt psychologists, many of whom emigrated to the United States during the 1930s, clashed with American behaviorists, who were deemed to hold overly mechanistic theories of human behavior. As for Köhler's book, McCulloch said, "in spite of his [Köhler's] endless searching, you will be convinced that he has not found the place of value in the world of fact." For McCulloch, locating "value" in a world of fact and potentially solving the mind-body problem was only possible by accounting for purposeful behavior through negative feedback mechanisms like the ones recently discussed by Rosenblueth, Wiener, and Bigelow. McCulloch resented the "existing hiatus" between mental terminology and physical terminology, but saw that a bridge was being made by showing that the "properties of systems which are like our nervous system necessarily show those aspects of behavior that make us call it 'mental'—namely, ideas and purposes."

Although psychosomatic research had a strong presence in the Chicago area and also integrated study of the brain and psyche, McCulloch had little patience for this approach. As McCulloch wrote to Gerty in November 1943, he supported the pursuit of biological explanations of mental conditions and felt his own work to be directly relevant to psychosomatics. However, he felt left out of Alexander's psychosomatic enterprise in Chicago, having only recently been invited to participate with a "pseudomembership." To McCulloch's dismay,

[t]he group is dominated by a clique of psychoanalysts centering around [Flanders] Dunbar, who is a vivacious but uncertain quantity. I have attended several of its meetings and addressed one of them. They all suffer from this fundamental defect: namely, that they have somehow split medicine into psyche and soma, and are unable to make the hyphen stick. Mentally, the bulk of the members are still struggling with the Descartian [sic] split between mind and body—which is so abhorrent to my way of looking at these problems that I feel frequently at a loss in talking to them. It is my inclination at all times, therefore, to steer clear of any entangling alliances, for

to my mind all problems of psychiatry necessarily involve psyche and soma as two aspects of one problem—nothing more. ...[100]

McCulloch was expressing here a disdain for not only psychoanalysis but also the slowly growing influence of psychoanalytic ideas on American psychiatry. McCulloch saw the mind and the functionally organized brain were one and the same.

McCulloch and Pitts had presented their early ideas to Rashevsky's seminar in January 1943.[101] University of Chicago physiologist Ralph Lillie was there, and he wrote to McCulloch with reactions to the paper by McCulloch and his "remarkably talented young associate."[102] Lillie found their assertion of the unequivocal "all-or-none" activity of neurons and thus the applicability of "either-or" logic quite interesting—akin to quantum theory—however his problem with the theory was the question of how activities and systems in nature are best represented to the human mind. While "logical and mathematical models," Lillie told McCulloch, "are admirable for most departments of physics," all too often physicists forget that they are simply *models*. In Lillie's view, the logical account of nature includes "only a part of its reality, and, I am convinced, a small part at that!" While logical principles are constant by definition, in nature, entities are deemed constant via observation and experiment. Here, Lillie was questioning and qualifying not only the applicability of logic to natural science, and the relation of logic to the natural world, but the attribution of "reality" to logical and mathematical models. While he admitted that logic is modeled after nature, it still is a human invention. Logic applies to the natural world as long as it "clearly and unambiguously conveys experimental facts."

McCulloch replied congenially, said that he agreed with most all of what Lillie had pointed out to him, and explained what he and Pitts had in mind, which transformed their neurophysiological account of mind or psyche into a naturalization of logic and epistemology.[103] This was an embodiment of mind: McCulloch believed his work with Pitts showed that we can for the first time see the activities of the mind as "the natural consequence of an organism ... without the introduction of a metaphysical hybrid like the ergotic hypothesis." McCulloch pointed out to Lillie that the observed regularity—the all-or-none neurons—bears a one-to-one correspondence with psychic atoms or "psychons." He conceded that "neither our logical description nor out actual neurones include all reality," and was pessimistic that scientists would ever be able to know anything about the natural world except some "abstraction from reality." But nevertheless he and Pitts focused on the regularities. The development of a physiological theory of knowledge would have profound implications for understanding the human psyche, both healthy and diseased.

## 4.7    Conclusions

Through varied scientific practices—laboratory work on the brain, performing the role of neuropsychiatrist, acting as an egalitarian mentor, McCulloch's scientific path had branched into two forks—the basis of his scientific life being his laboratory work as a neuropsychiatrist in the basement laboratory of the INI and his more formal, model-building practices with Pitts. His performance of the identity of egalitarian mentor captures an important way in which his personal and professional lives intersected and its legacy for cybernetics. His openness and ability to collaborate with junior scientists allowed a key facet of cybernetic practice to come into being. His laboratory work and modeling practices came together in the rhetorical ways that McCulloch promoted his own brand of scientific psychiatry. Previous accounts of the McCulloch–Pitts collaboration have simply cast their work as a "precursor" to cybernetics and AI, or else as a straightforward computational theory of mind. Yet has several historians of the human sciences have observed, the mind-brain problem, which strikes one at first glance as a philosophical issue, has much to do with disciplinary relations, the ways that science, medicine, and philosophy are related, and how the natural and human sciences intersect.[104] In essence, pursuit of a scientific account of the mind boils down to the question of what role science should play in psychiatry and what a scientific psychiatry should look like. There were many answers debated during the 1930s and early 1940s. For McCulloch, objective accounts of the mind would be mechanistic, materialistic, and grounded in propositional logic.

   McCulloch's unabashedly materialist picture of the mind rejected the psychosomatic split between psyche and soma and portrayed feedback mechanisms as the key to understanding the mind and its diseases. By elaborating hypothetical arrangements of idealized neurons that incorporated negative feedback, McCulloch and Pitts argued that one could come up with neurophysiological mechanisms that could explain mental activity—both its "formal" (logical) and "final" (purposeful) aspects.[105] McCulloch viewed the methods of abstraction and idealization as powerful tools, and the best way to embody mental and psychological processes in the physiology of the central nervous system. This allowed McCulloch, in his own estimation, to present a scientific account of the mind and its activity that was mechanistic and required no immaterial principle.[106] Although McCulloch's identities had multiplied during the early 1940s, through his associations with Pitts, Lettvin, and Rashevsky, we cannot treat his modeling practices as distinct from his laboratory practices as a neuro-psychiatrist. In fact, his role as an egalitarian mentor to Pitts and others facilitated an enormous transformation in McCulloch as a scientist. It generated a research program in theorizing and modeling the nervous system—a pursuit which had previously been non-existent or at least peripheral to his day-to-day activities.

McCulloch's stance on the mind was also tied to his lack of respect for disciplinary boundaries and domains. Moving fluidly between disciplines, professionally, personally, and intellectually, by 1942 McCulloch was able to integrate his clinically motivated studies of cerebral localization and his philosophical interest in logic and the mind. As we shall see in chapter 5, this integration continued as he moved further into the domain of biological psychiatry.

# 5  The Neuropsychiatrist

## 5.1  Introduction

As McCulloch's identity as a neurophysiologist was transformed by his practices at the Illinois Neuropsychiatric Institute (INI) in Chicago during the 1940s, he acquired the related but distinct identity of a neuropsychiatrist. Although he still did neurophysiology research at the INI, his laboratory had a firm place within the Department of Psychiatry. Now a Professor of Psychiatry in the medical school, and ultimately director of the Research Laboratories at the INI, McCulloch took on new administrative and institutional roles, and entered a new, more stable phase in his career.

The identity "neuropsychiatrist" was more than a professional label or a signifier of more seniority. It also does some work in reflecting McCulloch's rather idiosyncratic way of practicing brain research in the context of clinical psychiatry. The diverse scientific projects that occupied McCulloch between 1944 and 1952 can be organized into three distinct but related groups. The first set of activities, which generated the most publications, concerned the functional organization of the cerebral cortex. Using macaques, chimpanzees, and cats as experimental animals, McCulloch's collaborators were neuroanatomists Gerhardt von Bonin and Hugh W. Garol, neurologist Percival Bailey, and neurophysiologist Horace W. Magoun. The second group of practices involved brain research that was directly relevant for clinical contexts, including analyses of EEG records, carbohydrate metabolism, schizophrenia, epilepsy, and chemical warfare.[1] His collaborators on these projects included psychophysiologist Chester Darrow, biological psychiatrist Ladislas J. Meduna, neurosurgeon Arthur A. Ward, and psychoanalyst Franz Alexander.

Finally, especially as the Macy Conferences were underway, McCulloch began to engage more with theoretical practices that generated models of the brain's functioning in light of three related developments: the Rosenblueth-Wiener-Bigelow work on negative feedback, his own work with Pitts on the logic of neural relations, and an emergent conception of the brain as an information processor—an idea that was spreading through the nascent cybernetics group rapidly.[2] These models integrated his interests in scientific foundations and models of the brain and mind, and more often than not, McCulloch discussed these modeling practices in publications and lectures outside his traditional disciplinary home. While at first glance this work appears as a straightforward mechanization of mental activity—both healthy and diseased—we must consider McCulloch's activities within the context of his new institutional milieu. If prior to this period, McCulloch's interest in scientific foundations and philosophy were peripheral to his laboratory practices on the brain, by now, as he performed the dual identities of neuropsychiatrist and cybernetician, his unique practices of theoretical modeling intersected with his empirical concerns about the diseased brain and its organization. Although McCulloch also performed the identity of a cybernetician during this period, a process that will be examined in detail in the next chapter, by the late 1940s, his practices both as a neuropsychiatrist and as a cybernetician began to intersect, almost to a point where they became indistinguishable. Examining such practices more closely reveals both a resonance and tension in his intellectual pursuits. It is hard to imagine how his more theoretical, abstract, yet precise approach to the functional organization of the brain could be reconciled with his work on Meduna, which was organized around a much more vague framework in clinical pathology; and even more so how this stood in relation to his ultimate attack on psychoanalysis, despite his association with Franz Alexander and psychosomatics. Ultimately McCulloch's identity of "neuropsychiatrist" captures the complex relations that existed between laboratory research, neurophysiology, clinical research, and psychiatry, and the diverse, almost disparate projects that occupied him.

**Figure 5.1**
Warren and Rook McCulloch, Old Lyme, 1944. Courtesy the McCulloch Family.

## 5.2   American Psychiatry and Neuropsychiatry during World War II and the Postwar Period

What kind of discipline was American psychiatry, and how can we situate McCulloch within it? The eclecticism that had defined American psychiatry during the Depression continued during the 1940s, yet overall, the discipline expanded in unprecedented ways. Soon after the US was pulled into the war after December 1941, the number of soldiers suffering from war neuroses—debilitating anxiety attacks, depression, tremors, amnesia, and recurrent nightmares—rose exponentially.[3] The wartime work of psychoanalysts William C. Menninger, Roy G. Grinker, and John P. Spiegel transformed American psychiatry and pushed the field toward a rather eclectic psychodynamic model, although somatic conceptions did not disappear entirely.[4] In 1945, Grinker and Spiegel published their widely influential book *War Neuroses*. Within this new psychoanalytic framework, mental breakdown was no longer understood as stemming from a predisposition to mental illness, but rather as the reaction of normal people to extraordinarily stressful situations.[5] Many psychiatrists of the 1940s, following on the reforms of Adolf Meyer and others, constructed their identity as practitioners dealing with the "whole person"—an individual's thoughts, behaviors, worries, and passions in direct relation to their environment.[6] American psychiatrists following the war were brimming with confidence: mentally unwell soldiers during the war had been successfully subjected to short-term psychotherapy in the psychoanalytic framework.[7]

Yet the 1930s and 1940s also saw the rise of enthusiasm for the scientific promise of somatic therapies such as fever therapy, insulin therapy, prefrontal lobotomy, and electroconvulsive therapy (ECT).[8] As psychiatric practice moved out of the milieu of the mental hospital, psychiatrists attempted to cast their practices in the language of what was seen as a successful scientific medicine. Joel T. Braslow has called this a new "therapeutic rationale" for understanding disease and its treatment and cure.[9] Prior to this period, particularly in the age of the asylum, psychiatric disease was often understood as a behavorial disorder—particularly in the cases of severe mental illness such as schizophrenia.[10] In many cases following the war, within this new therapeutic rationale, mental illness was understood as a treatable biological disease. These therapies offered hope to desperate practitioners who previously had no effective way of treating patients suffering from severe conditions such as manic depression. Further, psychiatrists who embraced the medical, scientific model of psychiatry were critical of "talk therapies" and found some level of scientific rigor in somatic treatments.[11] Yet, like the other somatic treatments of the era, insulin shock therapy had no theoretical basis, and along with other problems with somatic treatments—in particular the ethical ramifications of "experimenting" with new treatments on chronically ill patients—led to controversy and attacks from more dynamically oriented psychiatrists.[12]

Neuropsychiatry emerged during this period as a distinct medical specialty, and although the term itself dates back to the context of World War I, its origin seems more political than medical or scientific.[13] According to psychiatrist Frankwood E. Williams, a prominent figure in the American mental hygiene movement, he and neurologist-psychiatrist Pearce Bailey needed to come up with a term in the context of the Committee for Organizing Psychiatric Units for Base Hospitals during World War I. As both neurologists and psychiatrists would be involved in such units, a name was needed that more accurately reflected the work being done and satisfied the need for professional identity on the part of workers at the units. They decided on "neuropsychiatry" to appease both neurologists and psychiatrists. Years later, neuropsychiatry became a diffuse term. It could simply refer to the field concerned with behavioral disturbances that clearly had an organic basis and organic effects, such as psychoneuroses.[14] In the context of military psychiatry, neuropsychiatry involved psychoanalytic practices aimed at treating war neuroses.[15]

At the Illinois Neuropsychiatric Institute, neuropsychiatry simply meant an institutional connection between neurology and psychiatry, and an interdisciplinary ideal. By the fall of 1944, INI director Francis J. Gerty requested a grant-in-aid from the Rockefeller Foundation for support of work in the Psychiatric Division of the INI.[16] A 1950 report reinforced Alan Gregg's interdisciplinary vision of psychiatry, broadly construed to include neuroanatomy, neurophysiology, neurochemistry, clinical neurology, neurosurgery, and clinical psychology.[17] Gerty sought support for psychiatric *research*—he felt that the investigation of psychiatric problems required the use of chemists, physicists, physiologists, psychologists, and sociologists, where basic and clinical research were closely aligned. Teaching and clinical work took second place as Gerty was determined to assemble a stellar research staff.[18] At the INI, boundaries between neurology and psychiatry and between basic and clinical research were fluid.

When McCulloch began at the INI he was beginning to shed the role of junior scientist. By August 1944 he was head of the Neurophysiological Laboratory and had more administrative duties. By the mid-1940s, Taffy and George were in their mid-twenties, David and Mary Jean were teenagers. As his younger children were entering their adolescent and teen years, McCulloch's life became more stable as he acquired a secure institutional home. With his directorship of the Research Laboratories, he had more administrative duties but at the same time more room to be bold.

Around the time that McCulloch and Pitts published their paper on the logic of neural networks, McCulloch had already begun to enter into other collaborative relationships that drew him further into the worlds of biological psychiatry. Jerry Lettvin had moved to Boston to take up an internship in neurology at Boston City Hospital. Pitts had remained in Chicago until the autumn of 1943, when he moved to Cambridge, registered as a student at MIT, and began working under the supervision of Norbert Wiener.[19] Yet the McCulloch household—both in Hinsdale and Old Lyme—was still full

**Figure 5.2**
McCulloch (front row, fifth from left), Rook McCulloch (to McCulloch's right), and colleagues in front of the Illinois Neuropsychiatric Institute, 912 S. Wood Street, Chicago, circa 1940s. Ladislas J. Meduna is in the front row, second from the right. McCulloch Papers, Scrapbook "From the Basement Research Laboratories, Neuropsychiatric Institute, Chicago 6/26/52," Box 3, Series VIII Photographs, WSM Papers, APS.

of activity and guests. The environment was free-spirited and open. Politically, while not a "hard boiled political advocate" of any party, McCulloch was a liberal Democrat. For his daughter Mary Jean, this meant that he was "open-minded and a fair judge of character and of what was going on in the world."[20] This openness translated into the diverse collaborative projects that occupied McCulloch at the INI.

McCulloch's interactions with members of the nascent cybernetics group would not begin to be regular until 1946. In 1944, the INI's

Psychiatric Division occupied the large basement laboratory. The Department of Psychiatry was organized into three divisions: general psychiatric, psychological, and neurophysiological (under McCulloch). Staff also included Franz Alexander (who divided his time between the INI and the Institute for Psychoanalysis), Ladislas Meduna (since 1943), and Frederick A. Gibbs, an authority on electroencephalography. Although support of brain research by the federal government began to eclipse private support by the end of the 1940s, McCulloch's work in this lab was supported by the Josiah Macy Jr. Foundation and the Rockefeller Foundation, who would ultimately supply over $245,000 in grant money to the INI. Like the Rockefeller Foundation, the Macy Foundation had been motivated by interdisciplinary concerns since its inception.[21] Officially established in 1930 to assist "scientific investigations of the fundamental aspects of health, of sickness, and of the methods for the relief of suffering,"[22] the Foundation's mandate was to foster a "multi-professional" approach to medicine and an integration of medicine with the biological and social sciences.

## 5.4 McCulloch's Practices as a Neuropsychiatrist

Within a year of arriving at the INI, McCulloch cited the uniqueness of the institute in a request to the Department of Welfare for more support for staff, citing the INI as one of only three state-sponsored places in the US where basic scientific research was closely aligned with clinical studies of mental patients (the other two being Worcester State Hospital and the New York Psychiatric Institute).[23] McCulloch's practices at the INI were diverse, both in form and content—so diverse that his scientific output here poses a challenge for anyone attempting to categorize the practices or make sense of them as a coherent whole. Yet a few rough lines can be drawn.

First, most of what occupied McCulloch during this period was the functional organization of the cerebral cortex. Here he collaborated with Bailey, Garol, von Bonin, and Magoun, and worked mainly with chimps and macaques.[24] By 1944, McCulloch was considered such an authority on the subject of cortical organization that he was asked to compose an essay for *Physiological Reviews*.[25] Also in this year, after he lured electroencephalographer Frederic Gibbs to the INI from Harvard, McCulloch pushed his work on functional organization toward interpreting EEG records, with help

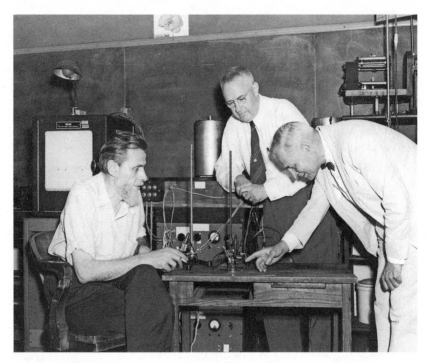

**Figure 5.3**
In the research laboratory at the INI: McCulloch and two of his closest collaborators at Illinois: Percival Bailey and Gerhardt von Bonin, 1948. Courtesy the McCulloch Family.

from Chester Darrow.[26] Eventually, McCulloch used his localization technique to study epilepsy and Parkinsonian tremor.[27]

The second set of McCulloch's laboratory practices had more direct clinical relevance and dealt with the biochemical, neurophysiological, and therapeutic connections between carbohydrate metabolism, fatigue, schizophrenia, and epilepsy. McCulloch was invited to a two-day Macy-sponsored conference on war neuroses held in January 1944, organized at the suggestion of the Office of the Air Surgeon. The meeting addressed the use of hypnosis and narcosis in the psychotherapy of war neuroses—two methods (one psychodynamic, the other somatic) that were associated with the wider project in neuropsychiatry by Grinker and Spiegel.[28] Soon after this conference, McCulloch and analyst Franz Alexander (who had trained Karl Menninger) approached Frank Fremont-Smith and the Macy Foundation for support for a project on the "influence of spontaneous goal-directed interest, or "zest," and its absence in the physiology of fatigue."[29] With Alexander, McCulloch studied patients suffering from fatigue and "vegetative retreat."[30] Both ultimately were seen by McCulloch as part of a larger common theme of "carbohydrate metabolism" in mental disease.

In June 1943, McCulloch related to Fremont-Smith the results of work he'd been doing on major psychoses, particularly schizophrenia, and requested funds from the Macy Foundation.[31] Hungarian neurologist Ladislas Meduna had arrived at the INI that summer, and McCulloch began working with him on the biochemical and metabolic aspects of mental conditions such as schizophrenia, as part of a search for physical and chemical therapies of mental disease.[32] Insulin therapy, pioneered by Viennese doctor Manfred Sakel in 1928, involved injecting a large dose of insulin into the patient, which generated a severe hypoglycemic state and insulin "shock," usually in the form of a coma, convulsions, and a full brain seizure. Once revived with sugar, the patient's mental condition apparently improved. In 1936, Meduna had presented a variation of Sakel's insulin shock therapy by using the drug metrazol, which induced convulsions in schizophrenic patients.[33] His theoretical basis for this treatment was his erroneous assumption of a "biological antagonism" between epilepsy and schizophrenia: that patients with epilepsy never developed schizophrenia. Ultimately through the 1930s, Meduna's therapy, along with Sakel's, became widespread in use, along with electroconvulsive therapy (ECT), which appeared later and involved passing a current through the brain, which generated an epileptic seizure in the patient. By 1940, nearly every mental hospital in the country had introduced insulin and metrazol therapy.[34]

Building on his work with metrazol therapy, Meduna had been studying the role of carbohydrate metabolism and the isolation of a hormone-like substance in the urine of schizophrenic patients, as well as working at the Manteno State Hospital, a psychiatric hospital in Kanakee County, Illinois. Meduna and McCulloch characterized schizophrenia as a biochemical condition, and wanted to contribute to understanding of the metabolic processes associated with mental disorders and psychoses.[35] With support from the Macy Foundation, they focused on carbohydrate metabolism in schizophrenic patients.[36] When studying a group of schizophrenic patients, they took samples of blood from the patients, injected the samples into rabbits, and then studied the effects of this "schizophrenic blood" on the rabbits. In 60 percent of cases, the injected blood somehow opposed the action of insulin, and thus the rabbits were unable to process sugar at a normal rate. Thus, Meduna concluded, schizophrenic patients differ widely in their blood chemistry. This work led Meduna and McCulloch to distinguish between two groups of patients. The "true" schizophrenics (who do not produce the insulin-opposing factors), have certain characteristic traits: they have dissociative symptoms, "lack of any strong feeling," yet a "completely normal sensory system."[37] In contrast, the patients whose systems *did* produce the anti-insulin factors had many sensory problems—they are unable to see or hear accurately and plagued by delusions and hallucinations. Meduna and McCulloch concluded that this group had a separate disorder that they named "oneirophrenia," from the Greek *oneiros*, meaning dream. These patients, besides being dream-ridden and subject to hallucinations, also had faulty sugar metabolism; schizophrenics, on the other hand, had irrational associations of ideas, emotional "coldness," and no defect in sugar metabolism. Thus the insulin treatments had markedly different effects on each group. Insulin and shock therapies were used to treat such patients, with oneirophrenics requiring a higher quantity of insulin.

Meduna and McCulloch saw their work on carbohydrate metabolism as having wide implications for understanding psychoses more generally, and for making sense of the theoretical basis for metrazol therapy and ECT. As McCulloch reported to Fremont-Smith in June 1948:

By using oneirophrenics that have spontaneous remissions or that regularly relapse when "cured" by electric shock, we have become convinced that both on going into a psychosis and coming out of it the change in insulin resistance precedes the change in mental function by at least 12 but not more than 48 hours. Puerperal, postpartum, and some manic-depressive psychoses clearly belong in our oneirophrenic group, in that they are resistant to insulin and, in our experience, no case that we have tested and found not resistant has been certainly "cured" by electric shock. This poses the question, "How can electric shock alter carbohydrate metabolism?"[38]

McCulloch speculated it may have been a consequence of chemical alternations in the carbohydrate metabolism in the brain. In his work with Meduna, McCulloch interpreted psychiatric disorders such as schizophrenia from a physical-biological standpoint—as Meduna wrote to Alan Gregg in 1949, they felt that "psychoneurotic conditions are physical conditions which can be cured by physical means."[39] For McCulloch, physiological and biochemical knowledge of the brain could direct clinical practice, as it would provide a scientific and theoretical basis for somatic therapies.

This perspective directed another facet of McCulloch's work at the INI: to study the state of the brain during the seizures induced by convulsive therapy—whether it be initiated by electric shock, metrazol, or another convulsive drug.[40] By simultaneously recording blood flow, pH, oxygen pressure and other factors during such a seizure, McCulloch and his colleagues were able to "determine the instantaneous state of the brain during the convulsion" and analyze it biochemically for carbohydrate metabolism, essentially getting a picture of what the treatment actually does to the patient's brain.[41] McCulloch related this to experiments he had done for the Medical Division, Chemical Corps, US Army on "chemical warfare," which investigated the action of sodium cyanide on the brain—convulsions and cerebral death. However, in the fifty cases they had treated, the psychoses had not been altered "in any way whatsoever."[42] Related to this work and Gibbs's work on various forms of epilepsy (grand mal, which involves violent convulsions; petit mal, where the patient has a sudden lapse into unconsciousness; and psychomotor epilepsy, characterized by irrational behavior), Meduna developed carbon dioxide treatment as an alternative therapeutic for various psychiatric disorders.[43] The treatment involved inhaling carbon dioxide, and intitally gained widespread appeal, as it seemed to allieviate the symptoms of neurotic patients, particularly obsessive ruminations.[44]

Guided by Meduna's work, McCulloch also studied the problem of relieving catatonic schizophrenic patients by having them inhale large quantities of carbon dioxide, and found that this treatment had been given to a series of psychoneurotics as a control, who had their symptoms permanently relieved. McCulloch also received support from the Office of Naval Research for "Investigation of Changes in Hyperoxic Convulsions" from June 1946 through to December 1951, essentially support for work in carbohydrate metabolism, but also to study the physiological basis of Meduna's carbon dioxide treatment.[45] McCulloch reported to Fremont-Smith that their next "major engagement" is with the United States Navy: "a study of those seizures which occur at about three atmospheres straight oxygen and a little lower if some carbon dioxide be added." They had built an "animal tank" to pursue this study. McCulloch reported to Rockefeller officer Robert S. Morison in September 1946 that injections of sodium cyanide depress activity in the cerebral cortex without causing irreversible neuronal changes—a significant difference from the production of "anoxia" by other methods.[46] This was all part of a project to understand metabolism and blood flow not only in the normal brain, but especially in the brain of patients with psychoses, in an effort to treat the psychoses using biochemical methods. Ultimately Meduna's carbon dioxide treatment failed to produce consistent results.[47]

McCulloch's identity as a neuropsychiatrist was tied up with a trend wherein the cure of behaviorial problems was located in the brain and body. Some of these somatic therapies have been critiqued as barbaric and their scientific efficacy questioned—by our own standards of treatment, only ECT has survived from this period as an acceptable form of somatic therapy.[48] Yet such critiques miss the interesting question of why such treatments were embraced in the first place. As historians, we need to make sense of the contradiction between our revulsion at such invasive mutilating therapies and the historical actors' use of and enthusiasm for them.[49] To do so, we must understand the history of psychiatry's attempts to achieve scientific status. As Braslow has observed, psychiatric therapies embody particular forms of knowledge and scientific practices.[50] In part too, such therapies may have surged because the postwar period saw mental illness emerge as a national problem. In 1948, a *New York Times* editorial on "New Horizons in Psychiatry" pointed to mental illness as the "number one" health problem for the foreseeable future: half of all hospital beds were occupied by mental patients.[51]

The variety of projects that occupied McCulloch, and his administrative roles, did not always put him in a good light. For example, Morison reported in February 1947 that at a dinner, McCulloch "talked a good deal in his usual way and I was sorry to find that on repeated acquaintance he tends to repeat himself more and more dogmatically. Unfortunately there is no one at Illinois capable of forcing his manifest abilities and energies in a constant direction for more than a few minutes at a time."[52] Mrs. McCulloch, Morison observed, seemed

tired and depressed, and I gather that both of them are worried about their two children who are not proceeding very happily through the storms of adolescence. Also in their house are Stanley Gerr's two children who are waiting out their parents' present marital difficulties, plus Norbert Wiener's 13-year-old daughter who apparently got into a jam at MIT and is working in McC's laboratory where she seems much happier. ... Incidentally, there are two or three other people in McC's outfit who have obvious difficulties of one sort or another, and it seems clear that he tends to pick his staff largely on the basis of their need for help. Although one can't help admiring his very warm heart, it seems doubtful if McC. can assemble an effective group in this manner.[53]

McCulloch's personal warmth and inclusivity was reflected in the variety of projects that defined his scientific life at the INI. Gerty saw all of McCulloch's work during this period contributing to the main goal of the lab, which for Gerty was "a better understanding of the grave constitutional psychoses which are the chief burden of mental institutions."

## 5.4   The Theoretical Foundations of Psychiatry

Sometime in the late 1940s, McCulloch listed as one of his chief scientific contributions, in addition to his work on the functional organization of the cerebral cortex, "theoretical work on psychological consequences of neurophysiological laws."[54] As early as 1944, he began employing the theoretical style of his 1943 work with Pitts to making sense of the nervous system. At this point, this was still a parallel, peripheral scientific activity. In a communication to Fremont-Smith, McCulloch sent a paper that "embodies one of the two central notions which are always in the back of my head in insisting that the Society of Psychosomatic Medicine needs a counterpart in somatopsychic medicine."[55] He continued:

My previous theoretical paper with Pitts indicated in what sense the formal, or logical properties of mentality are consequences of the go-nogo nature of nervous activity whereas this one [The Heterarchy of Values] turns the topologist loose on the value philosophers, for it indicates the relations to be expected in the realm of purposive behavior from the anatomical complexities of the nervous net mediating feed-back.

McCulloch wanted Fremont-Smith's permission to state that this work had been done under a grant form the Macy Foundation, since, McCulloch wrote, the work would not have been possible if Macy funding had not freed him from stressful, intense laboratory work. A close look at some of McCulloch's reflections on the state of psychiatry and brain research helps make sense of this move.

McCulloch and Meduna were both very committed to an "organic" view of mental disease.[56] McCulloch was somewhat sympathetic with psychosomatic researchers, however he often expressed frustration with their approach. In remarks given at a June 1944 meeting of the American Society for Research on Psychosomatic Problems, following a paper by Flanders Dunbar and George Soule entitled "Suggestions Preliminary to a Psychosomatic Nosology," McCulloch expressed discomfort with the term "psychosomatic"; in his view Dunbar and Soule had

> made clear that they do not believe in any theory of mind apart from body. That they bother to make this so clear indicates at once that they are trying to lay a ghost—the very ghost that is responsible for the prefix "psycho-somatic," when that hoary and unholy ghost is laid we can omit the prefix and say "medicine."[57]

McCulloch said that psychosomatic researchers had simply perpetuated Cartesian dualism, in which mind simply "paralleled" the body or soma. With its association with psychodynamics—which deals exclusively with "ideas and purposes," psyche and soma are forever hyphenated, and the "ghost" is not laid. To do so, one must show how ideas and purposes are "manifestations of the soma—i.e., of efficient cause and extensions into time and space." McCulloch referred to the mathematical tools he'd presented for doing so in his publication with Pitts, as well as the work that Wiener and Rosenblueth had done on the use of negative feedback to explain purposeful behavior. McCulloch jauntily concluded that today "it is fairly clear that the mental aspects of life can be treated as specific statistical aspects of the reactions of nerve cells."

McCulloch felt he was responding to the failure of certain other pursuits to unite physiology and psychiatry. Psychobiology, psychoanalysis, psychology, neurophysiology had all failed to measure physical or chemical changes in the brain that explain bodily changes and symptoms in the mentally ill: "The challenge to physiology is to construct, for psychosomatic medicine, a theory, right if possible, and if not, then demonstrably wrong ... I expect to fail."[58] The "subconscious" doesn't do it, Köhler's Gestalt psychology failed to do it, and Sherrington, McCulloch lamented, concluded that "Mind goes more ghostly than a ghost." The way to go about addressing the problem, for McCulloch, was twofold: first to study the nervous system from a biological standpoint—its anatomy, its metabolism, its circuitry; and secondly, to use a mathematical and theoretical framework. McCulloch then swiftly moved to his symbolism for describing the logical relations between neurons based on the all-or-none law, which, for him, demonstrated the "fundamental formal properties of mind." When changes in thresholds between neurons occur, "illusions, confusions, hallucinations, delirium, stupor or death" results. McCulloch believed his analysis of the "formal" aspects of mentality allowed one to claim that mind goes "less ghostly than a ghost. It is but circuit action of a relay net of which the blue print is incomplete, and which has some loose leads to be soldered in by success and suffering, or, in the long run, by use and wont."[59]

For McCulloch, theoretical models based on the logical relations of neurons and negative feedback mechanisms fed directly into discussions of mental disease. The relationship between his neurophysiological theory and psychoanalytic hypotheses was to place limits on them, "in the same sense that physics does on engineering."[60] Neurophysiological theory would serve as a fundamental basis for all of the myriad specific cases encountered by the psychiatrist. At the end of his remarks, McCulloch considered Meduna's oneirophrenia, whose symptoms are related to insulin resistance. The metabolic effects and symptoms can be relieved by convulsive therapy, such as electroshock. However, McCulloch remarked,

[w]e simply do not know how shock works. We have several reasonable hypotheses to check, all based on the fundamental certainty that unless carbohydrate metabolism of brain is normal its circuit action can not be normal. Physiology has, therefore, given us a theory which serves as a matrix of fairly well verified propositions from which to deduce hypotheses. Let us argue from it boldly and experiment unhesitatingly. Then if the theory is wrong we will find it out the sooner. The brain remains the Nodus Gordius of interconnected chains of all-or-none links. The surgeon may cut it if he knows where. The physician may dissolve its rust and prevent corruption if he knows with what. It remains for the psychoanalyst to untie it if he knows how. From the bottom of my heart I wish him God speed.[61]

The most promise in this pursuit could be found in the work of the nascent cyberneticians. Around the time of this talk, McCulloch wrote to Fremont-Smith (following a conversation), that he felt that the

group which is forming around Pitts and Wiener ... stands a better chance of contributing to breeching [sic] the woeful breeches between physiology and psychology and physiology and sociology than any lone worker anywhere. Nevertheless, I am plowing ahead on theory, usually at night, and I believe I have just begun to see the light with respect to the origin of hypotheses. ...[62]

McCulloch continued to participate in Rashevsky's seminar even after Pitts had moved to Cambridge. By the end of the decade, McCulloch's addresses to various interdisciplinary groups from psychiatrists to electrical engineers began to place his theoretical modeling practices with Pitts in a broad scientific framework the incorporated philosophy, the humanities, psychiatry, and communication.

Indeed, by 1945, seemingly in parallel with his laboratory practices, McCulloch began generating theoretical papers that integrated knowledge of the central nervous system, psychiatric questions, and the running cybernetic analogy of brains and computing machines. In the main these were lectures, often published in non-neurophysiological or psychiatric periodicals. Several were collaborations with Pitts. These reflections constituted a scientific practice that at once was a promotion of the cybernetic vision, a defense of the worth of theory in studies of the brain, and a contribution to biological psychiatry. For example, McCulloch's review essay on the functional organization of the cerebral cortex interweaved his extensive technical knowledge on laboratory work in neurophysiology with more recent work stemming from the group coalescing around himself and Wiener.[63] Noting that the most important experimental practices in functional organization came from work on primates, particularly macaques and chimpanzees, McCulloch presented the cortex as governed by reflexive actions, conceived as negative feedback: "to the extent to which its central nervous components can be influenced from elsewhere, notably the cerebrum, it can be made to seek other states, or, to use an engineer's term, be reduced to a servo-mechanism. Nothing but the name is new."[64] Departing from this analogy, McCulloch rejected other metaphors for making sense of the brain's function: Hughlings Jackson's accordion, Walshe's juke-box.[65] In this review, McCulloch took stock of historical and recent work on the cortex but at the same time explored the question of how best to conceptualize the cortex's organization. McCulloch presented the brain not simply as an object of laboratory investigation but as the seat of mental and neurological disorder—he gestured to the cortical mechanisms leading to *petit mal* seizures as well as mechanisms of reverberation that could account for memory. He explored the number of neurons in the cortex and the logical possibilities for connections. Finally, he referred to his own theoretical models devised with Pitts on the perception of universals and gestured to Rosenblueth's and Wiener's work on purposive behavior. More than simply a metaphor, such models are "fancy" which "cleaves to fact, prescribes experiment, predicts outcome, invites refutation."[66] They were not, for McCulloch, purely hypothetical.

A 1948 address presented a grand narrative of the history of thought, and promoted the cybernetic view whereby teleological mechanisms based on negative feedback could account for purposeful behavior across disciplinary boundaries.[67] While McCulloch tackled the brain head-on, he presented no precise theoretical model. Rather, he engaged in rhetorical practices that elevate the cybernetic style. Although such models had originated during the war, and had been "incorporated into certain engines of death and destruction," there is enormous potential for applying them to understanding human beings for better ends:

I know of no utopian dream that would be nearer to everybody's wishes, including my own, than that man should learn to construct for the whole world a society with sufficient inverse feedback to prevent another and perhaps last holocaust. There may at least be time for us to learn to recognize and decrease the gain in those reverberating circuits that build up to open aggression. We cannot begin too soon.[68]

A year later, to an audience of psychiatrists, McCulloch applied the cybernetic method of generating hypothetical mechanisms based on negative feedback to understanding psychoneuroses. McCulloch presented the main problem in his typical humanistic fashion:

I am inclined to sympathize with the good Saint Thomas Aquinas, that patient ox of Sicily, who lost his temper and died of the ensuing damage to his brain, when he met in Siger of Brabant the arch-advocate of two incompatible truths. In his time the schism was between revelation and reason, as in ours it is between psychology and physiology in the understanding of disease called "mental."[69]

Traditional neurophysiology has failed to bridge this gap, and the "alternative tradition"—in McCulloch's words, "Freud's trichotomy of the soul"—is no better: "both accept as real the separation of body and mind, although there is none in nature." For McCulloch, the only bridge between this apparent gap is "the science of signals, newly developed out of the art of communication." McCulloch went on to display his mastery of neurophysiological knowledge and how characterizing the functioning of the central nervous system in terms of servomechanisms would have potential for accounting for causalgia (burning pain) and vegetative retreat.

That same year, McCulloch presented an address to the American Institute of Electrical Engineers, in which he presented the brain as a logical machine with ten billion relays, or neurons.[70] Here he recast the therapeutic technique of ECT and studies of brain metabolism in terms of the computing machine model. Neuroses, in this case, occur when negative feedback goes "regenerative"—which fixes behavior in ill-adaptive ways. He also moved toward general functions of the nervous system such as memory and visual perception.

For the McCulloch of the late 1940s, the worth of the cybernetic framework was in its enormous potential for generating a sound theoretical foundation for understanding the nervous system—in both its healthy and diseased states. While theoretical modeling was an important aspect of McCulloch's scientific practices during this period, his activities were all translated via his identity as a neuropsychiatrist. In part, McCulloch fashioned himself as an opponent of the dominance of psychoanalysis in American cultural and academic life.

Psychoanalyst William C. Menninger, who had led the Neuropsychiatry Division of the Surgeon General's Office during World War II, promoted a medicalized version of psychoanalysis in the postwar era, and his efforts led to the psychoanalytic framework dominating American psychiatry and psychoanalysis itself becoming a cultural phenomenon.[71] Following the war, Menninger and other advocates for the expansion of psychiatry observed the rise of mental illness in American society more broadly, as hospitals were filled with mental patients.[72] While Menninger's role in this process was to in fact dilute Freudian psychodynamic theory and distance himself from the "extreme," "metaphysical" flavor of pure psychoanalysis, psychoanalysis thrived on multiple fronts.[73] In a 1952 speech to the Chicago Literary Club (in which McCulloch had been active since at least 1942), entitled "The Past of a Delusion"—a satirical jab at Freud's 1927 lecture "The Future of an Illusion"—McCulloch attacked psychoanalysis as unscientific:

For thirty years I have been among psychoanalysts, and held my tongue, and kept the peace. Not wittingly have I told one of them, analyst or analysand, what I have said tonight. Take my advice. Say nothing to them. Read their scriptures; listen to their lectures if you will; but say nothing! If you prove them wrong in anything they will explain it away, as they have always done, with one more hypothesis ad hoc. Delusions defend themselves that way. They are impervious to logic and to fact. If that were all you would just waste your breath.[74]

Earlier in the speech, he had likened psychoanalysis to a sect:

Psychoanalysts say that they have discovered that gold is a symbol for feces, but they formed a sect in psychiatry where there should be none. This sect so controls the teaching hospitals of our city that no one may be a resident of psychiatry unless he is approached by them for membership in their sect. To become a member he must be psychoanalyzed, for which the analyzed must pay the analyst who took the Hippocratic oath. ...[75]

McCulloch was not constructing a dichotomy in American psychiatry— between "talk therapy" and somatic treatments—but rather attacking the "tribal" nature of the growing faction of American psychoanalysts. Not every proponent of somatic treatments was as critical of psychoanalysis. Jonathan Sadowsky has shown that interpreting the history of American psychiatry through the metaphor of a pendulum, whereby psychiatry has swung back and forth between two extremes—one emphasizing psyche and the other soma—misses the very nuanced, pragmatic, and eclectic dimensions of the field.[76] Such a historiographic framework exaggerates dogmatism and obscures the many continuities in psychiatric styles during the middle of the twentieth century. In fact, as Sadowsky shows, there were many shared assumptions between advocates of psychodynamic therapies such as psychoanalysis and proponents of somatic treatments such as electroconvulsive therapy. While it is true that there were several prominent critics of ECT, many psychodynamic psychiatrists worked within a framework that allowed for a biological interpretation of mental disease. Freud himself was trained in neurology and had assumptions that were more biological than one might first suspect.[77] The response of American psychoanalysts to ECT was also not monolithic, and just as it was not a given that analysts would be hostile to biological frameworks, researchers in somatic therapies did not exclusively reject psychodynamic treatments. Overall, American psychiatry continued to be defined by eclecticism and a ubiquity of "mixed styles."

Thus, McCulloch's critique of Freudians was not only scientific and intellectual but also sociological—based on a perceived dogmatism in the field. This is telling not only as an instance of McCulloch's rhetoric but also highlights a tension in the various scientific projects that occupied him at the INI.

## 5.5 Conclusions

Standard accounts of the cybernetic project have associated the movement with the discontinuities that were seen to define the pre- and post-war periods in American science, particularly in terms of patterns of patronage and the conceptual frameworks through which investigators made sense of what it is to be human.[78] While I acknowledge such discontinuities, I believe the story is more complex. Firstly, through the lens of McCulloch's story, we see that many of the patrons for the activities and practices of the cybernetics movement aimed to promote a better understanding of the human brain and its role in mental illness—philanthropic projects that had already been in place for decades. Indeed, until the early 1950s, McCulloch's primary patrons were the Rockefeller and Macy Foundations. Further, the concerns with scientific foundations and scientific theory that preoccupied McCulloch during the 1930s at Yale evolved during the 1940s into practices that integrated empirical knowledge of the brain with a conceptual framework that reduced functions of the mind to logical relations.

McCulloch began his work at Illinois in 1941, fresh from his work with Dusser de Barenne and his exposure to a community of scientists engaged in a search for foundations—a junior scientist in desperate need of an institutional home. By the end of the 1940s, having just turned fifty, McCulloch was described by one RF officer as:

a slender, lithe, bearded figure, who leaps to the blackboard to diagram what he is telling you, who sees the human nervous system as a living net of 10,000,000,000 automatic relays, the biological analogue of an electronic calculator, and who for relaxation writes sonnets. His present preoccupation is brain chemistry. In particular, Dr. McCulloch and his associates are seeking to unravel the process by which the brain burns sugar to energize its thinking and feeling.[79]

Clearly McCulloch's eleven-year stint at Illinois witnessed multifaceted shifts in his identity as a scientist and as a person. The beard had grown greyish-white, his children all but grown and presumably out of the house (his eldest, Taffy turned 26 in 1952 and his youngest, Mary Jean, turned 22). Personally and professionally, he had entered a stage in his life where his intellectual pursuits had multiplied. An administrator, host to less-grounded colleagues and students, promoter of laboratory work on the functional organization of the brain, biological psychiatrist, theoretician—McCulloch the neuropsychiatrist at once performed diverse roles, speaking to diverse audiences and patrons. Yet at the same time, his need to the brain's functioning and development of foundations began to intersect. How did this happen?

First, there was more room for theoretical work. Such modeling practices began on the periphery of McCulloch's scientific life, yet by the late 1940s began to take center stage. Indeed, in his theoretical publications between 1949 and 1952 while at Illinois, psychiatry or psychiatric problems are not mentioned, and he generated very few publications in the field.[80] Second, McCulloch began to take more comfort in making rhetorical statements about the cybernetic project that involved grand sweeping claims about the relations between science, philosophy, history, and medicine. In an assessment of McCulloch for the Teagle Foundation in July 1947, Morison found it difficult to give a definitive assessment since he considered McCulloch a "warm personal friend."[81] He described McCulloch as "one of the most complicated personalities now doing research in the physiology of the nervous system. ... His work with Dusser de Barenne is considered top quality. Since arriving at the INI McCulloch has had to spend more time on administrative duties, however he has continued to publish a few papers on 'diverse subjects,' which are more difficult to evaluate." Morison attributed this to the fact that McCulloch had "an unusually original mind and is constantly trying to develop new approaches to old and difficult problems ... like all outstandingly original people, Dr. McCulloch probably makes some errors in judgment from time to time, and you may find that more conservative physiologists do not always immediately accept his more extreme statements." Initially such flourishes were received with some polite hesitation. However, as we shall see in chapter 6, ultimately McCulloch's deliberately "loose" language—a key feature of his evolving identity as a scientist—challenged traditional modes of practice and presented barriers in spreading the cybernetic vision across disciplinary boundaries.

# 6  The Cybernetician

## 6.1  Introduction

Through his connections to Macy Foundation officer Frank Fremont-Smith, Warren McCulloch became the primary architect of the Macy-sponsored meetings on cybernetics, both as organizer and as the movement's most colorful spokesperson. From the outset, he wanted the meetings to focus on mathematical and theoretical practices in studies of the central nervous system. An undated document generated sometime in early 1946 outlined his ambitious plans for a meeting to bring together mathematicians, engineers, physicists, and brain researchers, with McCulloch as chair. The gathering would take place at the Illinois Neuropsychiatric Institute, and in addition to people from far and wide geographically, McCulloch wanted a broad disciplinary distribution. It was quite a tall order:

Disciplines of mathematics to be represented by mathematicians familiar with problems of communication engineering or electrical engineers or physicists with proper mathematical background and no one to be included from this group who is not interested in the central nervous system;

Neurophysiology to be represented by neurophysiologists familiar with the electrical approach to nervous activity and with an adequate knowledge of mathematics, physics, and the general problems of communication;

at least 1 man from the discipline of muscular or neuro-muscular electrophysiology;

at least 1 physical chemist familiar with the problems of the central nervous system;

perhaps 1 biochemist familiar with the problems of the central nervous system;

at least 1 psychologist familiar with the first two disciplines;

preferably 1 neuropsychiatrist with similar qualifications;

1 anatomist familiar with electrophysiology.[1]

Norbert Wiener, Walter Pitts, and John von Neumann were on the tentative list as representatives of mathematics, with Nicolas Rashevsky for mathematical biophysics. Electrophysiologists included George H. Bishop and Rafael Lorente de Nó; Ralph W. Gerard and Arturo Rosenblueth appeared on the list as experts in the central nervous system. Given the emphasis on mathematics, McCulloch clearly aimed for an extension of his work with Pitts, which had generated hypothetical mathematical models of the central nervous system.

While it is not clear what became of this planned meeting, or whether it indeed took place, the sketch is a first iteration of McCulloch's vision for the Macy conferences: such interdisciplinary configurations came to characterize the meetings, which were held from 1946 to 1953.[2] While the Rosenblueth-Wiener-Bigelow work on negative feedback had generated enthusiasm in the diverse group that had attended the Cerebral Inhibition meeting in 1942, following a second meeting at Princeton in January 1945 (where mathematician John Von Neumann spoke on computing machines and Lorente de Nó gave a joint presentation with McCulloch on the problem of organization in the brain) there seemed to be consensus, according to Wiener, that "the subject embracing both the engineering and neurology aspects is essentially one, and we should go ahead with plans to embody these ideas in a permanent program of research."[3] Wiener's own vision included plans, following the war, for a formal society and journal, and a research center either at MIT or elsewhere. The Rockefeller Foundation was interested, and, Wiener happily told Rosenblueth, "McCulloch and von Neumann are very slick organizers. ..."[4]

The story of cybernetics—both its origins and its supposedly unified vision—has been told repeatedly, both by historians of and participants in the movement.[5] Often the cybernetic story is interpreted either from the vantage point of the emergence of artificial intelligence and cognitive sciences, or from a very Wiener-centric position that highlights the military, "control" facets of the cybernetic project. This emphasis on Wiener is not surprising, given his role as a key communicator of the cybernetic "brand" in his 1948 book *Cybernetics, Or Control and Communication in the Animal and Machine*, which galvanized the movement and captured the popular imagination during the postwar period. Yet despite McCulloch's and Wiener's reconstructions of the cybernetic story,[6] cybernetics was much more diverse and dis-unified than is commonly understood.

If we place McCulloch more centrally in this story, a different image of American cybernetics begins to emerge. A close look at McCulloch's role in generating enthusiasm for the meetings, his origin stories, his activities at the meetings, and his application of the cybernetic framework reveals that cybernetics for McCulloch was very much about understanding the brain and mind. Taken together with his work as a neuropsychiatrist, which overlapped with his participation in the cybernetics movement, we see that cybernetics ultimately allowed McCulloch to do two things. First, it became the framework through which he elaborated his ideas about the theoretical foundations of psychiatry, as he continued reject the Freudian perspective and promoted a biological foundation for psychiatric research and practice. Second, it provided him an intellectual and institutional space (in the figurative sense) to cast his own scientific project as a philosophical one—the pursuit of a scientific (i.e., biological, mechanical) theory of the mind. These two endeavors were two sides of the same coin.

A McCulloch-centered focus also provides new revelations about the cybernetic project itself. His list called for a gathering of scientific disciplines and, at first glance, it strikes one as an interdisciplinary project. Indeed, some scholars have labeled cybernetics an "inter-discipline."[7] The Macy-sponsored meetings on cybernetics took place against a backdrop of postwar discussions about the roles of science, technology, and medicine in American society, cast in the language of interdisciplinarity. Macy Foundation medical director Frank Fremont-Smith touted the cybernetics meetings as exemplars of cross-disciplinary exchange. The alliances that characterized cybernetics epitomized a disciplinary dynamic that had come to characterize American sciences in the mid-twentieth century: the intellectual migration of physicists into the life sciences.[8] As table 6.1 shows, a significant portion of the core members of the cybernetics group came from mathematics or engineering disciplines. Wiener, Rosenblueth, and McCulloch aimed to apply practices from these exact sciences—often theoretical and mathematical modeling—to problems in the life and human sciences, and they were explicit and self-conscious about these practices. Disciplines—whether they be considered inter- or transdisciplines—are not disembodied sets of ideas but are defined by practices.[9] Such tools can modify disciplines and be key to their emergence and development.[10] Theoretical modeling was just such a tool for cybernetics.

McCulloch's and Wiener's aspirations also reflected strong connections between American cybernetics and the American unity of science movement, which also strove for interdisciplinary collaboration, albeit with a similar emphasis on the primacy of physics, engineering, and mathematics. In this context, fluidity was promoted between biology, the physical sciences, mathematics, and philosophy.

However, while the Macy Foundation initially painted a picture of the conferences as an extension of an interdisciplinary ideal that characterized postwar medical and human sciences, this interdisciplinary unity proved challenging to achieve.[11] In part this was due to the various meanings of "interdisciplinary" that were understood by participants. While Fremont-Smith saw interdisciplinarity as an egalitarian collaboration between life and human scientists on the one hand and mathematicians and physicists

**Table 6.1**
The core members of the original cybernetics group. Adapted from Heims, *The Cybernetics Group*, 285.

| Disciplinary Background | Members |
| --- | --- |
| Mathematics/Engineering | Norbert Wiener |
| | Walter Pitts |
| | Claude Shannon |
| | Julian H. Bigelow |
| | Leonard Savage |
| | John von Neumann |
| Social Sciences/Sociology | Gregory Bateson |
| | Lawrence K. Frank |
| | Paul Lazarsfeld |
| Psychology | Molly Harrower |
| | Kurt Lewin |
| | Heinrich Klüver |
| Neurophysiology/Physiology | Arturo Rosenblueth |
| | Rafael Lorente de Nó |
| | Ralph W. Gerard |
| Psychiatry/Neuropsychiatry | Warren S. McCulloch |
| | Lawrence S. Kubie |
| Philosophy | Filmer S. C. Northrop |
| Ecology | George E. Hutchinson |
| Anthropology | Margaret Mead |
| Neuroanatomy | Gerhardt von Bonin |
| Medicine | Frank Fremont Smith |

on the other, both McCulloch and Wiener strove not simply for collaboration but colonization. While interdisciplinarity can evoke images of unity, synthesis, and convergence, it is a term that is often conflated with multidisciplinarity and transdisciplinarity. Cyberneticians were after more than simply interdisciplinary communication. In their goal of applying mathematical and theoretical practices to problems in the life and human sciences, they were striving for *transdisciplinarity*—seeking to lay over a new way of looking at the world that would transform scientific practice across fields. Transdisciplinary projects have broad conceptual scope—they involve "breaking through disciplinary barriers,"[12] and often involve a single theoretical perspective applied to a variety of disciplines.[13] As a result, McCulloch and other members of the group repeatedly encountered difficulties in communicating their vision.

I illustrate this feature of cybernetics through a focus on McCulloch's cybernetic work: how he translated his work as a psychiatrist into the framework of cybernetics, his rhetorical strategies, and the nature of the barriers he faced. On the surface, participants from non-mathematical fields expressed frustration at their inability to follow the technical language of physics, engineering, and mathematics. Despite the strong collaborative rhetoric presented by the Macy Foundation, the conferences were marred by a lack of consensus and a lot of confusion. Notwithstanding their emphasis on unity, core members of the group, particularly McCulloch, deliberately employed what Leah Ceccarelli has called *polysemy*—a deliberate rhetorical strategy that involves employing concepts from different disciplines in an ambiguous way—a practice that ironically was a key feature of the cyberneticians' transdisciplinary rhetoric. Ceccarelli has articulated a specific variety of polysemy called *strategic ambiguity*—which allows one to accommodate different disciplinary groups and promote a unified view.[14] This was key to McCulloch's rhetoric. Finally, when McCulloch presented his work on the brain and perception as a potential solution to the mind-brain problem, critics attacked his theoretical, speculative style which emphasized generalities over empirical accuracy. In the end, responses to McCulloch's work reveal much about standards for making knowledge about the brain and mind and the values that particular standards reveal.

In part, my analysis is in line with Andrew Pickering's account of the practices of British cyberneticians W. Ross Ashby, W. Grey Walter, Gordon

Pask, Stafford Beer, Gregory Bateson, and R. D. Laing.[15] Like McCulloch, Ashby and Laing had backgrounds in psychiatry, and like McCulloch, all of the British cyberneticians had a strong commitment to understanding the inner workings of the devices they produced. Yet American cyberneticians differed from their British counterparts in that they had much more of a disciplinary agenda, and were much more interested in legitimacy as a scientific field. Pickering has referred to his British cyberneticians as anti-disciplinary, which perfectly captures the sense of playfulness and irreverence that the British group had. In terms of how they characterized the brain, British cyberneticians had an ontology—a conception of what sorts of objects populate the natural world and how they relate to one another—that was adaptive and performative, rather than predictive in the traditional sense. In contrast, American cybernetics was more strongly grounded in a deterministic, predictive framework.

## 6.2  Cybernetics, the Unity of Science, and Mathematical Biology

The context of World War II and the postwar period is crucial for under-standing the unity of science movement as it unfolded in America.[16] While prewar unity, as elaborated by Rudolf Carnap and Otto Neurath, was an effort to model philosophy on the sciences,[17] eliminate metaphysics, and present a unified language for science, postwar unity, as expressed by Har-vard physicist and philosopher Philipp Frank, was much more pragmatic. This reflected the local context of Cambridge, Massachusetts, where war-time collaborative projects were happening everywhere. Here the early meetings of Philipp Frank's "Inter-Scientific Discussion Group" took place, first held at Harvard in 1944.[18] These interdisciplinary meetings on scien-tific method consisted of expatriate members of the Vienna Circle, but also included Harvard and MIT scientists and philosophers who were sympa-thetic to the principles of the original group: Wiener and Rosenblueth had participated in both these meetings—for example, Wiener spoke at a Febru-ary 1945 meeting on "The Brain and the Computing Machine,"[19] and Rosenblueth had been at one of psychologist S. S. Stevens's earliest gather-ings in October 1940.[20] Among this group there was a "drive to interdisci-plinarity." The participants initially shared two principles of the logical positivists: unity of the sciences and physicalism.

The group evolved by 1947 into the International Institute for the Unity of Science, after 1949 located at the American Academy of Arts and Sciences in Boston, with help from the Rockefeller Foundation.[21] In part this dovetailed nicely with the vision behind Harvard's new General Education program, which aimed to break down disciplinary barriers,[22] fuelled by the excitement and promise generated by wartime collaborative projects.[23] Much of the rhetoric in Frank's early writings in this context lamented the specialization that had transformed twentieth-century sciences, which was seen as a "serious threat to our culture." He saw strong potential in interdisciplinary collaboration and its impact on liberal education.[24] Scientists and humanists needed to be united to allow the public to form a critical perspective on the sciences, and in the Cold War atomic era, with new alliances between science and the state and a perceived threat of Communism, the fate of western democracy depended on this.[25] Further, Frank's vision for unity incorporated consideration of the sociology of science as well as the logic of science, and aimed to engage with actual scientific practice.[26] Frank cited especially the fields of chemistry and physics, which in the context of the Manhattan Project had successfully merged, and the recent interactions between physics and biology and between science and the social and human sciences. While prewar discussions of unity had centered on language, Frank's new approach was about integrating scientific disciplines.[27]

In this context, in part, postwar unity was sought in cybernetic terms, and through Norbert Wiener's participation in the Institute for the Unity of Science, cybernetics became a key topic in discussions following the war. Indeed, many discussions took place at MIT's Department of Electrical Engineering. By 1952, in an operational dictionary developed by Frank and MIT social scientist Karl W. Deutsch, Communications Engineering and Theory emerged as a new category of analysis, with "message," "information, and "network" among the subcategories.[28] For his own part, Wiener had continually engaged in discussions on scientific method and the roles of practices characteristic of physics and engineering in the life sciences.[29] Wiener's wartime work and how it figured in the cybernetics movement has been well documented, both by Wiener himself and by historians.[30] A crucial element of this story is the relationship that Wiener had with one of his most important collaborators, Arturo Rosenblueth. The Rockefeller Foundation supported their work, which explored the connection between negative feedback and purposeful behavior in physiological cases such as reflexes, heart fibrillation, and homeostasis.[31]

Besides their collaboration in what they called "mathematical biology," a significant facet of Wiener's relationship with Rosenblueth was their shared stance on how best to generate scientific knowledge about complex systems. Reflecting their participation in the Inter-scientific Discussion Group, they had grand visions for the unity of science, and had even pondered the possibility of generating a treatise on scientific method.[32] In a 1945 essay on the role of models in science, they argued that scientists must employ abstractions and scientific models when faced with the complexity of natural systems and "high" order scientific questions.[33] Wiener's rhetoric about scientific practice and the work that he was doing with Rosenblueth was part of his cybernetic vision: to bring models and abstractions to bear on physiological problems. For Wiener, the disciplinary territory of this new emerging field straddled the boundaries between physiology and mathematics:[34]

Dr. Rosenblueth has always insisted that a proper exploration of these blank spaces on the map of science could only be made by a team of scientists, each a specialist in his own field but each possessing a thoroughly sound and trained acquaintance with the fields of his neighbors. ... The mathematician need not have the skill to conduct a physiological experiment, but he must have the skill to understand one, to criticize one, and to suggest one. The physiologist need not be able to prove a certain mathematical theorem, but he must be able to grasp its physiological significance and to tell the mathematician for what he should look.[35]

The collaborative efforts of Wiener and Rosenblueth, and their vision for the role of models in science figured prominently in the origin story for cybernetics developed by Wiener in his 1948 book of the same name: the book itself was presented as "the outcome, after more than a decade, of a program of work undertaken jointly with Dr. Arturo Rosenblueth. ..."[36] Wiener presented their collaboration as stemming from their shared belief that "the most fruitful areas for the growth of the sciences were those which had been neglected as a no-man's land between the various established fields."[37] This vision resonated with discussions of scientific unity. Many contemporary philosophers of science, such as Carl Hempel, R. B. Braithwaite, and Ernst Nagel, ultimately saw cybernetic explanations as a way of bridging the supposed gap between the physical sciences and the life sciences, and allowing for legitimate explanations of apparently goal-directed phenomena in terms of physico-chemical laws.[38]

Although Peter Galison rightly points out that it is impossible to understand the "Americanization" of unity without taking into account the military collaborations and applications in which many cyberneticians and unity of science advocates were involved, postwar American unity did retain some remnants of the prewar Carnapian emphasis on physics and a hierarchical view of the relations between the sciences. Further, at least in cybernetic rhetoric, postwar unity differed further in its emphasis on modeling as a key scientific practice (and less on linguistic unity). In many ways, cybernetics, with its language of universality, became what physics had been to the prewar unity of science figures like Carnap. Understanding purposeful systems in terms of negative feedback and the logic of computers would result in greater understanding across disciplines. Yet as we shall see, for various reasons, this unity was difficult to achieve.

### 6.3   Interdisciplinarity and the Macy Meetings: Official Rationale

Following the Second World War, the Macy Foundation decided to adopt a conference program as one of its primary activities—likely a result of the need to re-invent itself in light of recent increases in federal funds for medical research.[39] As a result, the foundation had more freedom to support new ideas, and less pressure to show immediate results.[40] The rhetoric used by the foundation in presenting the conference program was steeped in interdisciplinary language. In the face of increasing specialization, the rationale behind the postwar Conference Program was to increase opportunities for the open exchange of ideas and better communication and collaboration between scientific disciplines.[41] Traditional disciplinary boundaries, reinforced by institutional structures, could be overcome through the dynamic, unstructured interaction that could take place within the conference framework. Meetings were intended to be informal works-in-progress.[42] The foundation was optimistic that such a loose, open-ended atmosphere would generate a "constructive working relationship" between participants and opportunities for collaboration between scientists from varied fields.[43]

Heinz von Foerster once said, "[i]f not the word 'interdisciplinary,' then the concept may well have been invented by Frank Fremont-Smith," whom von Foerster described as "a gregarious elitist who set a trend in medical research by recognizing trend-setters and supplying them with the financial and organizational apparatus of his foundation. ..."[44] Fremont-Smith received his medical degree from Harvard in 1921, and taught neuro-pathology there from 1929 to 1936. He became Medical Director of the Macy Foundation in 1937 and held this position until 1960. In his intro-ductory remarks at the Sixth Meeting of the cybernetics group in March 1949, Fremont-Smith reiterated this interdisciplinary focus:

> The Foundation's interest in these conferences stems from its experience, over some years, with the problem of advancing research and from increasing recognition of the need to break down the walls between the disciplines and get interdisciplinary communication. This failure in communication between disciplines seems to be a major problem in every phase of science. ...[45]

Fremont-Smith called the cybernetics group the "wildest" of all his confer-ence groups, due to its disciplinary diversity. Communication was not easy, but Fremont-Smith still felt a timely and very pressing need for com-munication across disciplines. This was not motivated by any idealistic vision of scientific unity for its own sake, but rather a product of Fremont-Smith's post–World War II belief that the physical sciences had developed to such an extent that "social misuse" may result and "greatly delay any further progress in civilization."[46] Echoing the statements of Philipp Frank in the context of the unity of science movement, Fremont-Smith argued that communication across disciplines will help mitigate such misuse: "the physicists have given us the ultimate weapons of hostility. Now per-haps it is important for all of us, including the physicists, and the math-ematicians to learn something about the nature of hostility."[47] Recent developments in computing machines and goal-seeking devices, along with work in neurophysiology, will shed light on "the mechanisms under-lying self-correcting and purposeful behavior of individuals and of groups."[48] Again, the main axis, in Fremont-Smith's view, fell across the divide between mathematics and engineering on the one hand, and neu-rophysiology on the other. Indeed, the largest sub-groups in the cybernet-ics group were from physics, mathematics, engineering, and brain sciences [see table 6.1].

## 6.4   The Macy Meetings and McCulloch's Cybernetic Practices

Fremont-Smith and McCulloch were certainly on the same page in their transdisciplinary spirit, and the two became the real organizational forces behind these meetings.[49] The plan for the first Macy meeting was to have von Neumann speak on the formal behavior of computing machines, including memory, learning, and recording, and Lorente de Nó following up with "biological exemplifications." In the afternoon, Wiener would discuss goal-seeking devices and negative feedback, and Rosenblueth would speak on the biological implications of such devices. The evening of the first day would include anthropologist Gregory Bateson on theory in the social sciences and philosopher F. S. C. Northrop on theory and verification in the physical and social sciences. The second day was to address problems of psychology, psychiatry, and game theory. Fremont-Smith's rationale was to order topics so as to familiarize the group "with developments of theory and practice in computing machines and target-seeking devices and, in each case, indicate physiological exemplification." Following this the group

**Figure 6.1**
Attendees at the Tenth Macy Conference on Cybernetics (Princeton, NJ, April 1953). McCulloch is in the front row, fourth from the left. McCulloch Papers, Professional Photos, Box 3, Series VIII Photographs, WSM Papers, APS.

would discuss psychosomatic, psychological, psychiatric, and sociological problems that could be tackled within this framework.[50]

For his own part, McCulloch was clearly delighted at the prospect of interacting with mathematicians and philosophers, although he quickly understood the potential difficulties in communication:

> I am beginning to believe that there must be two meetings in time to come, one consisting almost exclusively of those interest [*sic*] in formulation of symbolic methods for dealing with problems, and the second concerning the application of these methods to the problems of psychology, psychiatry, anthropology, and sociology, paralleling it, in a sense, to agenda for the first and second day. I know that men in the latter fields are keenly aware of their want of theory and that many of them were fearful that men in mathematics and the physical sciences would be uninterested in tackling such nebulous problems as their own. ...[51]

McCulloch strove for foundations: a theoretical basis and a set of "symbolic methods" applicable to wide areas of the human sciences. Problems in these areas were "nebulous," and the rigor of mathematics and the physical sciences would remedy this. Theory would also help make the social sciences more rigorous, as he told Bateson and Margaret Mead, he wished to combat the habit of deeming the social sciences impossible to ground in theory, due to their historical rather than experimental nature.[52] Ultimately, the theoretical tools of wartime engineering and physics could transform scientific practice in biology, psychology, and sociology.[53]

Clearly, McCulloch's vision of interdisciplinarity was directional, and echoed Wiener's and Rosenblueth's views on the potential of modeling as a scientific practice. Lawrence K. Frank, who presided over a meeting at the New York Academy of Sciences on "teleological mechanisms" that included several of the core members of the cybernetics group, also stressed the value of the cybernetic framework: "the scientific endeavor has had a long and often bitter struggle to free itself from the animistic tradition in our culture. It is enough to refer to the mind-body-soul conception in religion and philosophy and the persistence of psychic entities and vitalistic assumptions in biology and psychology."[54] Frank also noted that the notions of purposive behavior and teleology had long been deemed to be associated with mysterious, supernatural "final causes," yet employing cybernetic practices could provide more effective methods for studying such purposeful systems.[55]

Like the Cerebral Inhibition meeting of 1942, these early gatherings generated intense enthusiasm among the participants, regardless of discipline. Throughout the late 1940s, the group continued to interact outside the Macy structure, with frequent family gatherings at the farm in Old Lyme, where Rook and Warren extended their openness and generosity to include the cybernetics group and their families. The summertime atmosphere at Old Lyme had always been free-wheeling and imbued with McCulloch's sense of "delight, fun," and "joy."[56] These meetings involved animated discussions. As David McCulloch recalled fondly, "Can you imagine how interesting a beginning we kids had with Warren, Walter Pitts, Jerry Lettvin, and a tribe of people from over the world who sat around our big oval dining table. I have never been able to capture that excitement."[57]

## 6.5   Second Macy Meeting and Beyond: Barriers and Disunity

Soon after the first official Macy Meeting on Cybernetics, the cybernetics group seemed to share a unified transdisciplinary vision. However, by the eve of the Second Macy meeting, small rifts began to develop. Frank wrote to McCulloch in October 1946 about themes he thought could fruitfully serve to organize the discussion at the Second Meeting:

1.   Biological feedbacks in order to foster the conceptual clarification that is needed and to avoid the taking over of a formula from the field of mechanics or electronics without recognizing what the biological feedback involves.
2.   A clarification of the concepts of signals, signs, and symbols, not merely in semantic or semiotic terminology, but in terms of the biological and cultural *processes* which are involved.
3.   The requirement of a mathematics for handling biological data as distinguished from the kind of mathematics that has been developed for classical physics, chemistry, and astronomy. ...[58]

Evidently there were linguistic and disciplinary barriers standing in the way of the cybernetic utopian vision: mathematics and modeling practices could not be applied to biological cases in a straightforward way. A lack of consensus on the meanings of key terms such as feedback also plagued the group, and more conceptual work was needed to bridge the gap between the formulae of physics and engineering on the one hand, and the biological systems under consideration. Further, more work was needed so that the non-mathematicians and non-engineers could make better sense of the relevance and usefulness of the engineering approach to problems in the life and human sciences.

Beyond the difficulties in applying mathematics to the life sciences, the obstacles faced by the cyberneticians stemmed from other issues. First, many outside the core duo of Wiener and McCulloch felt frustrated at the loose language that dominated discussions at the meetings. This lack of precision accompanied the transdisciplinary economy sought after by the group—Frank feared that more clarity was needed so that mathematical language would not "take over" biological formulations. Often terminology such as "purpose" or "goal" was used vaguely. Such polysemous language, although intended as a way of bridging disciplinary gaps, ironically had the effect of blocking consensus or unity. For example, the concepts of "field" and "Gestalt" were used repeatedly—and rather vaguely—at the first meeting; as McCulloch recalled: "[t]he word 'Gestalt' came up for clarification and it was at once apparent that five members of the group thought the remaining twenty were abusing the term."[59] Beyond this, the informal structure touted as a value by the Macy Foundation led to a confusing chorus of too many voices. By the time of the Third Macy Meeting, Rosenblueth wrote to McCulloch from Mexico, with his shared worry with Pitts that the next Macy meeting might "turn out to be another 'loose talk' festival, a danger which has been threatening us persistently from the start."[60]

Early spectators to the cybernetic endeavor were also critical of such intentionally loose language. Rockefeller Foundation medical officer and neurophysiologist Robert Morison, a friend and colleague of McCulloch's, had already expressed frustration at McCulloch's style of communication, which certainly must have colored his discussions at the early Macy meetings. After a 1946 dinner with several colleagues, Morison wrote that McCulloch

gave a prodigious display of versatility—simultaneously paying graceful attention to Mrs. S., reporting on his neurophysiological experiments, elaborating the mathematics of electrical circuit theory, discussing the classification of schizophrenia, and settling the family difficulties of various hardware salesmen who dropped by our table under the impression given by McC's beard that they had met him in the submarine service.[61]

A few months later Morison attended a lecture by McCulloch, which he deemed "apparently as clear as possible an account of the possibility of making a mathematical formulation of the function of the nervous system."[62] While Morison felt that McCulloch had given much time to the lecture, and felt it was a "worthwhile attempt," he couldn't follow all of it. One of McCulloch's assumptions was that the organization of neurons in the central the nervous system is entirely random, and when pressed on this point by Lorente de Nó, McCulloch had side-stepped the question. McCulloch's rhetorical style was connected to the analogy of the brain as a computing device, and Morison was skeptical of the analogy itself. In Morison's view, the model is weakened by the fact that "all models are merely analogies which may or may not tell us something about the structure of the biological system in which we are interested."

In winter 1947, Morison began to express further reservations about the cybernetic project. His concern was twofold: again he was frustrated at the style of communication that characterized not only McCulloch but the entire group, but this time he also cited his lack of disciplinary expertise in making sense of the mathematical models. The discussions at the Macy meetings and other gatherings were "rambling" and struck him as a "chaos of clear ideas. I had the curious impression of being able to follow almost every sentence without being able to make sense of the paragraphs."[63] Morison expressed a general puzzlement with the linguistic style of the cyberneticians—for example, brushing aside the relationship between purpose and final cause, and lumping together equilibria and goal-seeking mechanisms. Modestly, Morison admitted to Rosenblueth that a good deal of his difficulty "must be due to my own incapacity to follow the thinking of such experts as Wiener and yourself." However, it seemed that this was only a minor frustration. Clarity, for Morison, was a virtue in scientific practice that trumped anything else: "[l]ike it or not … a certain proportion of he efforts of creative thinkers has to be directed at making their thoughts intelligible to the less gifted people."

Morison's assessment of McCulloch may have been colored by his role as an officer from a rival funding agency, yet even participants were frustrated. With an air of humility that contrasted sharply with Wiener's and McCulloch's characters, Rosenblueth concurred with Morison's assessment and went so far as to say that he was reconsidering attending future meetings "unless a careful weeding of the group is carried out, because not only I but also Wiener, von Neumann, and Pitts have been quite irritated by the large amount of loose and confused talk which has taken place in them in the past."[64] Rosenblueth himself was apparently less carried away by cybernetics: talk of purpose and feedback was less important for him than straightforward work in what he viewed as "theoretical biology." Given Rosenblueth's remarks and omission of McCulloch, it is likely that McCulloch was the source of much of the "loose talk" he referred to. McCulloch's broad transdisciplinary perspective was tied to his speculative style.

On one level this disunity was personal—by spring 1947, a rift had emerged between McCulloch and Wiener over Pitts's apparent loss of a manuscript of Wiener's, and McCulloch's defense of Pitts.[65] Wiener, who had always been much more of an authoritarian mentor in contrast to McCulloch's more egalitarian style, cut off all ties to Pitts, Lettvin, and McCulloch. The split was worsened by a misunderstanding over an accusation by Barbara Wiener that McCulloch's "boys"—Pitts, Lettvin, and Oliver Selfridge—had seduced her in Chicago during the late 1940s, a charge that had more to do with Wiener's and his wife Margaret's territoriality over cybernetics.[66] Sadly, Pitts never recovered from Wiener's rejection and eventually became reclusive.

Disciplinary barriers also plagued the internal workings of the group. Psychologist Molly Harrower, who had been at the Fourth Meeting in October 1947, wrote to McCulloch about attending the Fifth Meeting on speech and communication in the Spring of 1948: "As you know, about fifty percent of the discussion is absolutely unintelligible to me, and while it is instructive to learn what I can provided I am not under any pressure and am simply an onlooker, the roles of the active and passive members of that group still bother me. …"[67] Harrower touched on the dynamic that characterized how *transdisciplinary*, not *interdisciplinary*, this group was. Cybernetics was not about a simple collaborative, egalitarian effort between members of the human and biological sciences on the one hand and those from the exact sciences on the other. Rather, the core group aimed to apply practices from the exact sciences to problems in the life and human sciences. Obstacles existed from both directions: life and human scientists found the language of physics, mathematics, and engineering difficult to understand; physicists and mathematicians failed to appreciate the complexities of living systems. At the Sixth Meeting in March 1949, anthropologist Margaret Mead expressed frustration in a characteristic remark that describes a clash of scientific cultures:

May I just say one thing? If we could somehow work it out, the reverse position in this group, so it would be in between the psychologists and the physicists, that would be fine. I think when somebody writes an equation on the board followed by more and more difficult ones, everybody in the room knows when they get left. There are some people like me who get left very soon, and there are some people who never get left. However, almost everybody in the room knows that that point in mathematics is one which I do not understand. I will not understand it in that language. I have to wait until somebody has said it in English, or with a different figure of speech, or has related it to my data before I can understand it.[68]

Despite the fact that Wiener's 1948 book *Cybernetics* had brought some level of unity to the group—they soon adopted his title as the new name for the Macy Conferences—by January 1949, McCulloch found unity at the meetings increasingly a challenge to achieve. While McCulloch felt that the "widest possible interdisciplinary thinking and communication" was the principal goal of the meetings, he did not believe a "real collaboration" had yet been realized at the Macy meetings.[69] In terms of the dynamics of meetings, even McCulloch even admitted it would have helped if speakers had been limited to no more than 15 minutes. Up until this point, the structure of the meetings was quite loose, with no fixed sequence of speakers.[70] As he told a Macy meeting of Chairmen and Editors, "[y]our troubles always come from somebody talking too long uninterrupted."[71] He suggested remedying this by including in the group figures whose problems "sit at the crossroads of the physical and social sciences" and for a general topic be chosen for each meeting, one that sits at such crossroads.

The Sixth Meeting, held in March 1949, seems to have been a turning point, particularly in McCulloch's efforts to address disciplinary barriers and generate more focused discussion. McCulloch reiterated the purpose of the meetings, which was to allow for the group to "talk together across the barriers between various scientific disciplines."[72] To this end, the focus of the meetings should be on problems of a kind and scope "that all of us are likely to have something to say or think about them." After a discussion with Fremont-Smith and others, McCulloch relayed that they concluded the topic should be more general, and suggested morphogenesis, memory, or language as a possible topic for the next meeting.

Despite McCulloch's efforts, a speculative "as if" style that Morison had targeted in McCulloch also defined the Macy meetings. Even John von Neumann, by late 1946, was pessimistic about the potential of the McCulloch–Pitts work for the study of neurological mechanisms themselves, since he found the model too general, which was an impediment to any real understanding of the human brain, "the most complicated subject under the sun—literally."[73] At the Seventh Macy Conference, in March 1950, Gerard critiqued the "digital approach" to the nervous system. He observed that when he reflected on the history of the meetings, the group had

started our discussions and sessions in the "as if" spirit. Everyone was delighted to express any idea that came into his mind, whether it seemed silly or certain or merely a stimulating guess that would affect someone else. We explored possibilities for all sorts of "ifs." Then, rather sharply it seemed to me, we began to talk in an "is" idiom. We were saying much the same things, but now saying them as if they were so.[74]

Gerard saw dire consequences for such recklessness, for members of the cybernetics group had legitimate internal and external responsibilities as scientists. Internally, each member was obliged to communicate as clearly and precisely as possible to each other. With the group being comprised of members from very diverse fields, Gerard noted "no one can be sure another's statements are facts or guesses unless the speaker is meticulous in labeling suggestions as such." As for the group's external responsibility, which Gerard felt was even greater, the group also must be sure to be clear communicators, and not give a "spurious certainty to a credulous audience."

## 6.6 "How We Know Universals" and "Why the Mind Is in the Head"

Despite McCulloch's efforts to overcome interdisciplinary barriers at the Macy meetings, his theoretical work on the nervous system continued to rile his colleagues, particularly brain researchers who had not taken up the cybernetic creed. McCulloch brought his brand of cybernetic practice to an external audience in his remarks at the Hixon Symposium on Cerebral Mechanisms in Behavior in September 1948, held at the California Institute of Technology. Nowhere are his rhetorical practices and philosophical aspirations more evident. Many historians view the symposium as a key event in the history of brain and behavior sciences, given its focus on bringing together psychologists and brain researchers and its apparent departure from the behaviorist ideal in psychology.[75] Most of the presenters at the

Hixon meeting were psychologists, with the exception of von Neumann, Lorente de Nó, and McCulloch.

While McCulloch's talk, entitled "Why the Mind Is in the Head," took place outside of the context of the Macy meetings, it reads as a manifesto of his cybernetic vision. The talk essentially presented the upshot of his 1947 work with Walter Pitts, entitled "How We Know Universals," which had applied their general theory of logical neural networks to the perception of auditory and visual forms. At the Hixon meeting, McCulloch recast this work within the rhetoric of the mind-brain problem.[76] The 1947 work had been a response to a challenge posed to the cybernetics group by the Gestalt psychologist Heinrich Klüver at the First Macy Conference in 1946, and had been influenced by the work of British psychologist Kenneth Craik and his mechanistic framework.[77] "Gestalten" referred to phenomena in perception where we do not perceived objects such as melodies, shapes, or the letter A, for example, as piecemeal fragments; rather such forms are experienced as a whole. Essentially, McCulloch's and Pitts's theory attempted to show that knowledge of the neuronal architecture of the cerebral cortex (the morphology and distribution of cell types) could be used to demonstrate how the perception of relations between parts of a form could remain constant despite changes of scale—in a sense how we perceive Gestalten. McCulloch and Pitts had reasoned that the image one perceives could be equated to a pattern of stimulus that is then subjected to dilatations and averaged over a group through mathematical transformations [figure 6.2]. For example, a triangular shape whose image is represented by a particular distribution of excitation in the cortex, would be perceived, through a transformation of a corresponding mathematical function, as the universal form "triangle," regardless of shape, size, or perspective. In a sense McCulloch and Pitts had turned Gestalt theory on its head—we perceive forms as experienced wholes because of what represented a complex, mechanical wiring diagram—of multiple parts. The whole—that is the phenomenon of perception as well as the perceived form—is the effect of many parts working together.

McCulloch and Pitts had invoked several pieces of evidence that they believed supported the theory: empirical neuroanatomical and neurophysiological data of Santiago Ramón y Cajal,[78] McCulloch's own laboratory work, and clinical findings. However, despite the fact that they presented empirical data, McCulloch and Pitts qualified the model as presenting a set

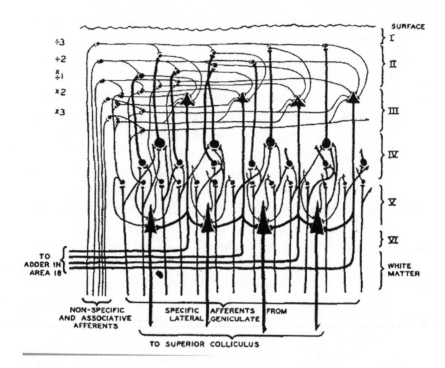

**Figure 6.2**

Pitts's and McCulloch's version of the visual cortex. Impulses relayed by the lateral geniculate from the eyes move down to layer IV, where they branch laterally and excite small cells singly and larger cells by summation. Large cells thus represent larger visual areas. From layer IV impulses go to higher layers. From there they converge on large cells of the third layer which relay impulses to the parastriate area 18. On their way down they contribute to summation on the large pyramids of layer V which relays them to the superior colliculus. From Pitts and McCulloch, "How We Know Universals," 134.

of "hypothetical mechanisms" that would be able to direct experimental and histological work.[79] In his Hixon presentation too, McCulloch almost continually referred to the theory as a *hypothesis*:

So far we have considered particular hypotheses of cortical function. They are almost certainly wrong at some point. Because they have already had to fit many disparate data, they are of little a priori probability. They prophesy the outcome of an infinite number of experiments, some of which are most certain to refute them.[80]

Clearly, a virtue of the model for McCulloch was its generality and potential to guide experimental investigation of the visual cortex, not its empirical accuracy.[81] The idealization of neurons as digital entities fed into McCulloch's conceptualization of the brain-mind as an information processor.[82] Within this framework, nervous impulses were cast, in McCulloch's terms, as "logical signals" or "atomic propositions on the move."[83]

Work in the history and philosophy science that deals with modeling has moved well beyond the traditional discussion of logical empiricist vs. semantic views about scientific theories.[84] Modeling as a scientific practice has been acknowledged as messy, diverse, and pragmatic.[85] Models can be seen as mediators between theory and the natural world, and as such they can be understood as somewhat autonomous.[86] According to Mary S. Morgan and Margaret Morrison, philosophically, we can make sense of models by inquiring how they are constructed, how they function, what they represent, and how they help us learn about the natural world. McCulloch's models of visual perception, and of neural activity in general, were constructed not simply out of data but were also driven by a theory—a set of assumptions about the nervous system and the brain. In terms of how his models functioned, for McCulloch, they were to guide experimental investigation, but also served as a means for a theoretical foundation—and scientific respectability—to neurophysiology and implicitly psychiatry. Thus McCulloch's models were tools: instruments of investigation. They were presented as malleable, open to alteration, and not absolutely representative of the central nervous system and its functioning. They were a means of intervention, not representation.

McCulloch's work, and the cybernetic practice in general, is a prime example of Alastair C. Crombie's scientific style "hypothetical modeling."[87] Mary S. Morgan has recently argued that to really make sense of models, we must understand the diversity of models as processes—as ways of "giving form" to ideas—rather than look for labels or disembodied categories to organize a taxonomy of models.[88] On Morgan's grouping, McCulloch was an idealizer—such modellers (often working in the field of physics) choose relations of interest and isolate them to produce simpler models of the world.[89] Yet McCulloch's model did not simply serve his modest aim of guiding neurophysiological experiment—he also spun his model, and the cybernetic project itself, as an attempt to bridge the "traditional gulf" between "mind and body."[90]

How novel was McCulloch's framing of his work in the language of the mind-brain problem? The term "mind" can take on a variety of meanings according to discipline. What "mind" means in psychology and even to different psychologists is quite different from what it means in philosophy or neuropsychology.[91] In fact, in postwar brain sciences, engaging with this question was not unusual, and discussions of mind cropped up everywhere. For example, the June 1951 meeting of the American Neurological Association held a *Symposium on Brain and Mind*,[92] and the decade saw several more meetings devoted to the connections between the brain, consciousness, and behavior.[93] All of these discussions had strong ties to research practices and recent work on how physiological and mental processes are organized. As for the reasons behind this postwar surge in addressing such brain-behavior connections, recent research had revealed the existence of a reticular system in the brain, which appeared to be responsible for regulating consciousness.[94] Neurologist Stanley Cobb in 1951 had observed that several scientific developments seemed to be at the root of what he saw to be a new focus on mind and treatment of it as a scientific object. He listed at least eight, including anatomical studies that revealed cerebral complexity, surgical work on conscious patients on localization of functions such as memory, detailed studies of neuron circuits,[95] and work related to the cybernetic project on memory, feedback, and scanning mechanisms for the recognition of universals.[96] All of these, Cobb argued, seemed to bring us "closer" to a better understanding of mind.[97]

In light of all this, how did Cobb define mind? He regarded consciousness as a component of mind, along with memory. However, he argued that the only tenable hypothesis for describing the mind is one that depicts it as a "changing dynamic mechanism." In his view, "[i]t is the integration itself, the relationship of one functioning part to another, which is mind and which causes the phenomenon of consciousness. … The brain is the organ of mind; its great complexity in man makes his thinking possible, but no study of the anatomy and physiology of one brain will ever explain mind."[98] Cobb defined mind as a question of convention, not a question of fact or truth.

Cobb's attitude was not unusual: for the most part, the notions of mind discussed by neurologists and neurophysiologists were quite generic. Often these discussions characterized mind as a "verbal cloak" for such processes such as perception, reasoning, learning, and memory.[99] While most brain researchers were happy entertain the possibility of a scientific theory of mind, as we saw in chapter 4, the generation that came of age in the 1930s and 1940s were largely working within a scientific and intellectual tradition established by the Nobel Prize–winning British neurophysiologist Charles Scott Sherrington. Sherrington held organization and the concept of *integration* to be key in understanding the central nervous system.

Yet rhetorically McCulloch found it important to position his work as addressing the mind-brain problem explicitly. This is an important aspect of McCulloch's rhetorical strategy. In his Hixon talk, McCulloch's rhetoric was ambiguous in several distinct ways. First, his language was constantly shifting between the discourse of science, engineering, and philosophy—a practice that was also prominent in his 1943 and 1947 publications with Pitts. For example, 1943's "A Logical Calculus" contains the abstract assertion: "To demonstrate existential consequences of known characters of neurons, any theoretically conceivable net embodying the possibility will serve."[100] In "How We Know Universals," the mechanisms of perception are translated into mathematical/engineering terms of input, output, and computation. Second, there was continual movement between discussion of "the mind" and functions such as "having ideas" (equivalent for McCulloch to "securing invariants"),[101] sensation, perception, memory, consciousness (having an "idea of ideas"), atomic propositions, and prediction.

### 6.6 Responses to McCulloch's Cybernetics

Responses to McCulloch's presentation and style varied. Generally, critics either attacked his mechanistic picture of the mind or his stance on the empirical accuracy of his models. Behind such reactions were disciplinary commitments and values about the role of science in understanding the mind and relations between the natural and human sciences. Certainly few neurophysiologists or psychologists would doubt that memory and perception are amenable to scientific study. Yet McCulloch rhetorically presented his project as having the potential to *solve* the mind-brain problem, which is much different than simply saying a certain restricted definition of mind can be tackled scientifically.

At the Cerebral Inhibition meeting, what generally troubled participants were McCulloch's loose language and simplifying assumptions. Neuropsychologist Karl Lashley said he was sympathetic to the ideas and general principles presented in McCulloch's talk, and was not entirely against the project of attempting to tackle the problem of the mind and brain by focusing on activities of neurons. However, he took issue with McCulloch's method, which stemmed from the computer analogy. In order to draw the analogy between the computer and the mind-brain, McCulloch needed to abstract specific properties from neural elements, some of which were speculative and hypothetical. Referring to McCulloch's model of visual perception, Lashley noted that the model involved a major oversimplification of the situation, and that "the behavior which is explained is behaviour which never occurs in the intact organism. It is a hypothetical behavior derived from assumptions of the system rather that the observed phenomena."[102] Lashley told the group he saw McCulloch's theory as too general: more specific phenomena would require a new set of assumptions, and thus his theory, which would have to be "revised to fit each special case," had little utility. Lashley also argued that McCulloch's theory involved assumptions that were unjustified by facts, particularly regarding neural structure and function, and that the idealized cortical structure ignores the fact that cortical structure varies anatomically between individuals. Lashley also attacked McCulloch's assumptions about the *nature* of neural activities involved in perception. For Lashley, theories of neural interaction must be framed in terms of *mass relations* of cells, and he pointed out that even the "simplest of behavior requires integrated action of millions of neurons."[103]

While McCulloch was willing to depart from empirical realities to push his theories, Lashley, in contrast, was more scientifically conservative and was unwilling to draw very many theoretical conclusions from empirical results.[104] It is arguable that this criticism might have been particular to Lashley: throughout his career, Lashley was critical of mechanistic analogies in studies of the brain, and engaged in a debate with Clark Hull in the 1930s and 1940s over the worth of machine analogies in understanding the human mind.[105] However, Lashley's remarks also reflect a general belief in the importance of experiment and observation in studies of the brain and behavior, or at least a certain primacy to empirical studies; and an appreciation of the uniqueness of individual brains, which challenged McCulloch's emphasis on generality. As Lashley addressed the meeting of the American Neurological Association in June 1951:

I am less impressed with the analogies between various machines and neural activity, such as are discussed in Cybernetics. … Descartes was impressed by the hydraulic figures in the royal gardens and developed a hydraulic theory of the action of the brain. We have since had telephone theories, electrical field theories, and now theories based on computing machines and automatic rudders.

I suggest we are more likely to find out about how the brain works by studying the brain itself and the phenomena of behavior than by indulging in far-fetched physical analogies. The similarities in such comparisons are the product of an oversimplification of the problems of behavior.[106]

According to Peter Cariani, Lashley's critique amounts to an "incrementalist intellectual conservatism" and a bias against theoretical modeling: it is more palatable to work toward smaller, incremental achievements rather than great leaps of insight about the big picture.

At the Hixon meeting, von Neumann also showed skepticism toward McCulloch's theory of memory based on reverberating circuits, noting that at the same time that McCulloch states he is giving a "sufficient" mechanism of memory—a "proof of possibility," he only raises the question of what the actual mechanisms are.[107] Wolfgang Köhler was critical of McCulloch's introduction of "ad hoc histological assumptions" which were not justified by the facts, and disliked McCulloch's habit of using psychological terms in a "diluted sense" and substituting one problem for another—having a goal is not the same as reaching a goal, for example.[108]

Reviews of the published version of the Hixon Symposium had similar things to say about McCulloch's presentation. Neurophysiologist John Eccles criticized the participants' general rejection of consciousness as relevant to the problem of brain and mind. He also had specific criticisms of McCulloch:

Professor McCulloch gave a characteristically racy discourse under the title "Why the mind is in the head." Undoubtedly McCulloch is in agreement with the main stream of neurophysiological thought in that he attempts to make the generation and transmission of impulses the dominating and controlling features in cerebral mechanisms. But too often his imagination and rich verbal imagery run away with the scientific argument. One can sympathise with Dr. Köhler's criticism that McCulloch introduces psychological terms into neurophysiology so that neurophysiological events with mistakenly be thought to have a psychological significance. The gulf between neurophysiology and psychology is speciously bridged. ...[109]

Eccles even wrote that he was dismayed that there was not enough criticism of some of the inaccurate statements in McCulloch's talk, such as those concerning the activity of photoreceptors in the eye.

At the Hixon Symposium, McCulloch found himself defending and clarifying his aims. Early on, he had suggested the key question for the group was to "ask the theoretical physicist to account for himself"—in essence, asking the theoretical physicist to account for his production of knowledge in theoretical physics.[110] McCulloch's choice of subject was not random:

Now, in asking that question, I did not ask that he be a psychologist, and I did not ask that he be a philosopher, and above all, I did not ask that he be a theologian. The entire problem, I am quite sure, will never be manageable if we begin at the most difficult end. I would much rather ask, merely, that it be a theory of the physicist, and my object in doing this is to put the question concerning *whether a machine of this order of complexity can state how it works* [my emphasis].[111]

McCulloch's approach reflected a common belief among core members of the cybernetics group: one can account for seemingly purposeful behavior using methods traditionally used in the physics and engineering sciences (such as mathematics and formal modeling) and that certain apparently "goal-directed" phenomena (such as mental activity) were *too complex* to be understood through empirical methods alone.

McCulloch defended his "quantization" of the activity of neurons by pointing to what he saw as the profound *complexity* of the cerebral cortex, which for him justified his theoretical and reductionistic approach: to understand in general how cortical areas are related, he argued, one should begin with hypotheses about particular areas. Disproof, McCulloch said, would only come with records of electrical activity in the cortex. In the discussion following John von Neumann's talk, McCulloch had emphasized what he viewed as the *explanatory goal* in studies of brain function. As he told the participants:

As I see it what we need first and foremost is not a correct theory, but some theory to start from, whereby we may hope to ask a question so that we'll get an answer, if only to the effect that our notion was entirely erroneous.[112]

One of McCulloch's most cherished philosophers was Charles S. Peirce. From Peirce, McCulloch adopted a particular stance on scientific reasoning that goes some way toward accounting for how he presented his models. Peirce had introduced *abduction* as a third form of scientific method in addition to induction and deduction. Abduction, for Peirce, involved studying facts and devising a theory to explain them"[113]; in essence, generating explanatory hypothesis that, once provisionally accepted, could then be subject to a test and either accepted or rejected.[114] Indeed,

McCulloch's theory of the perception of universals was eventually subjected to experimental testing and it was shown not to be accurate. Lashley ultimately criticized the notion of precise anatomical or structural patterns having functional significance, and later discussed histological evidence that contradicted their theory.[115] Years later, McCulloch reported that a member of his lab at the Illinois Neuropsychiatric Institute, Donald MacKay, had "constructed a square which could be made to balloon out, shrink down at any preassigned rate, or could be coupled back from one's own brain waves. According to all my notions, that square should have upset our ability to perceive form. It failed completely to do so. ... I think MacKay has thrown the biggest rock that can be thrown into my hypothesis."[116] MacKay had shown limits on the capacity of individual neurons to transmit information, undermining McCulloch's and Pitts's binary, digital model of neural activity.[117] But, judging from his remarks at the last Macy Conference on Cybernetics in April 1953, McCulloch was undeterred: his original aim was merely to come up with a hypothesis that could be subjected to disproof. As he said to the Macy group: "To insist on being wrong is to insist on there being something which can be checked. It is my notion that every scientific hypothesis has a reasonable expectation of being disproved. ..."[118]

## 6.7 Conclusions

McCulloch's scientific persona evolved in parallel with the emergence of cybernetics. By the end of the Macy meetings in 1953, his identity as a neuropsychiatrist had been eclipsed by his identity as a cybernetician. With this new identity, grand philosophical questions—those surrounding the mind-brain problem—came to the fore. McCulloch's model-solutions, with their associated language of negative feedback, were at once tools for achieving foundations (bringing coherence to the eclectic field of psychiatry) and also at the heart of a set of deliberately ambiguous rhetorical practices that in some ways proved to be an obstacle to transdisciplinary communication. McCulloch was invoking the term "mind" in a rhetorical way—to emphasize what he saw as the power of the cybernetic approach to allow a mechanistic (i.e., scientific) account of mind. Indeed, in a presentation at a 1953 conference on "The Validation of Scientific Theories," organized by the Institute for the Unity of Science in Boston, McCulloch claimed that "Cybernetics has helped to pull down the wall between the great world of physics and the ghetto of the mind."[119] In many ways, as McCulloch entered into discussions of mind during this period, his perspective had much to do with beliefs about the role of science and the relations between science, philosophy, and the humanities. Where traditionalists saw rigidity and firm boundaries between disciplines, McCulloch saw fluidity.

In a review of the published version of the Eighth Conference in 1952, MIT psychologist George A. Miller was highly skeptical of the cybernetic project.[120] With their obsession with "symbolic processes" across disciplinary divides, Miller felt the group was plagued by a "familiar dilemma"—the tension between a faction that compares psychological processes to machines (and ignores the complexity of real psychological processes), and the opposing faction that insists on the scientific (read experimental) method and is angered by "the vagueness and disagreement in the description of these complex, allegedly meta-mechanical phenomena." Miller was clearly in the latter camp:

Since the boundary between cybernetics and science fiction has never been overtly sharp, those of us seriously interested in the social and psychological applications of this kind of thinking have often wished for some standard of scientific respectability in this young discipline. The Macy conferences could supply such a standard, for the participants are uniformly men of solid achievement ... in their own fields. Instead of becoming a scientific standard-bearer, however, the conference has chosen to emphasize interdisciplinary propaganda. After eight conferences this line is running a little thin.[121]

Miller's remarks capture the main criticisms of the cybernetics movement that had been leveled by the early 1950s.

McCulloch's method went against the grain of epistemological standards in brain sciences. Regardless of what questions neurophysiologists or psychologists attempted to answer during this period, their method in obtaining an answer was by and large grounded in experiment: this was a methodological ideal. McCulloch's methodology, in contrast, "left the ground" so to speak and was based on formal modeling. For McCulloch, experimentation and empirical data played a secondary role in his initial explanations, and ultimately served to confirm or refute his theories. A consequence of this was what was perceived by some as a loose or inadequate relationship to empirical results or to the way that the nervous system actually functioned: the "entities" in McCulloch's mechanistic explanations were idealized and empirically impoverished.

Cyberneticians like McCulloch gave a particular status and weight to theoretical models. Philosopher of biology Sabina Leonelli has recently emphasized the diversity of modeling practices and their varied epistemological functions.[122] As Leonelli has observed, theoretical models may have explanatory power but lack empirical content, or vice versa—a model it seems should strike a balance between the two. Further, Leonelli suggests, models can possess other virtues, such as generality, tractability, and empirical accuracy.[123] Critics of McCulloch might not have been against models per se but may have had different epistemic goals for their models, or emphasized different virtues, like empirical accuracy.

When faced with criticism, McCulloch often presented his models as simply pictures of how the brain and mind might *possibly* function, rather than how they *actually* functioned, which flew in the face of scientific standards during this period. David Resnik has argued that "how-possibly" explanations generally lack empirical support, but can serve an important heuristic function by aiding the development of new theories and suggesting new areas of research.[124] Clearly McCulloch saw this as a virtue of his models. Models could open up possibilities—as Seymour Papert recalled with admiration, McCulloch would say to his detractors: "Don't bite my finger, look where I am pointing."[125]

For Wiener, for whom cybernetics was consistently about communication and control, the cybernetic story was weaved into the history of physics and engineering: the analogy between organisms and communication machines was grounded in their common ability to control entropy.[126] Yet eight of the non-exact sciences members of the cybernetics group were from brain and mind sciences. For his own part, McCulloch abhorred metaphysical language in discussions of mental activity (a throwback to prewar unity), but at the same time had his own part to play in military applications and ultimately, as he moved away from the discipline of psychiatry proper, became part of the postwar academic world of "interdisciplines" like cybernetics.[127]

McCulloch's move to Illinois was out of professional need and perhaps desperation; he took advantage of this new configuration of laboratory research on the brain and clinical treatment of patients in the INI. By the

1950s, given the success of psychotherapeutic methods in treating soldiers and veterans suffering from the psychological stresses of war, psychoanalysis became the dominant framework in American psychiatry.[128] By the time McCulloch had become a full-fledged cybernetician, and as his theoretical practices began to dominate his scientific life, he took advantage of new patrons in American science during the Cold War era and fled Illinois and the world of medicine to join Lettvin and Pitts at MIT.

# 7 The Engineer

## 7.1 Introduction

In May 1951, electrical engineer Jerome Wiesner, soon-to-be head of MIT's new Research Laboratory of Electronics (RLE), formally invited Warren McCulloch to join the lab as a research associate, officially to work on synaptic transmission.[1] Walter Pitts was already at MIT, Jerry Lettvin was soon to return to Boston from Manteno State Hospital outside of Chicago, and they were all delighted at the prospect of working together again.[2] McCulloch had earlier approached Wiesner and proposed the transfer of work he was doing at Illinois to MIT, since both he and Wiesner, who was in the communications sciences branch of the RLE, felt that "the facilities and support of an organization such as this were required to provide the extremely complicated instrumentation required for his work."[3] McCulloch and Rook arrived at MIT in October 1952, moved into 1 Shady Hill Square in Cambridge, and by December, Wiesner reported that while having McCulloch at the RLE was certainly an "experiment" for both parties, "[w]e are extremely satisfied with the way the thing is working out. ..."[4]

Given McCulloch's home in a neuropsychiatry unit at Illinois, this was more than a simple institutional migration—it was a shift to a dramatically different disciplinary world. The RLE was just one of the many large-scale, multidisciplinary units that the US government created and sponsored during the Cold War.[5] However, while the RLE had connections to the military and industry, it also had research climate built on stable funding, a degree of intellectual freedom, and cross-disciplinary collaboration. As was the case with many other federally-funded centers during this period, members of the RLE were encouraged to pursue whatever projects satisfied their intellectual curiosity.

It was precisely this open, fluid, yet stable scientific culture that drew McCulloch away from the world of neuropsychiatry and medicine at the University of Illinois. Obviously, on the surface, the promise of no teaching or administrative responsibilities was appealing, along with a reunion with his closest colleagues Pitts and Lettvin. In May 1952, McCulloch wrote to his brother Paul about his excitement at the RLE position, writing that it would

[r]elieve me of all executive chores, for which I am ill fitted, and leave me complete-ly free to devote my time to research on the central nervous system, with two collaborators with whom I worked happily and profitably over several years. ...[6]

However, this new institutional context provided more for McCulloch than freedom from administrative duties. McCulloch flourished in the open, multidisciplinary climate that the RLE provided. Besides work on synaptic transmission and the visual and auditory systems—experimental work that in fact owed much to Lettvin's efforts—McCulloch recast his interest in the functional organization of the brain in the theoretical framework of information theory, communications sciences, and engineer-ing, and the majority of his scientific practices focused on theoretical modeling.

McCulloch had already been engaging in interdisciplinary practices at Illinois, although up until this point, his main research focus was on prob-lems in neuropsychiatry, and involved hands-on laboratory research on the brain, motivated by clinical questions. His cybernetic work, although closely related, never received the same amount of attention or institu-tional support. With his arrival at the RLE, his modeling practices had an institutional home. At Illinois, McCulloch's research on the brain had sup-port from philanthropic foundations such as the Rockefeller Foundation and the Macy Foundation. At the RLE, McCulloch's position as research associate meant that he devote himself full time to research, taking advan-tage of massive amounts of funds being poured into the RLE by the US government. In turn, as his modeling and rhetorical practices came to cen-ter stage, the brain for McCulloch acquired a new identity as a scientific object. No longer an object simply of medical interest, (with theoretical analysis on the periphery), the brain, for McCulloch, now stood at the nexus of philosophy, laboratory neurophysiology, communications engi-neering, information theory, and organizational theory.

McCulloch began to perform the identity of an engineer, using mathematical and rational methods to solve a myriad of practical problems including reliability and cognition, and through these practices, his identity as a neuropsychiatrist was transformed. McCulloch's performance of the identity of engineer involved moving further into the realm of the practical and applied sciences. Always a hands-on scientist, when asked in a 1969 interview why he became an engineer, he said, "Oh very simply in the first place I wanted to find out what was *going* [*sic*] in the brain. I'm using that in the old sense."[7] Here McCulloch evoked nineteenth-century scientist James Clerk Maxwell, whose insatiable curiosity about the natural world led him to continually ask of anything that "moved, shone, or made a noise, 'What's the go o' that?' and if not satisfied with your answer, 'What's the particular go of that?'."[8] Among countless other scientific achievements, Maxwell had made contributions to modern control theory and his work directly contributed to the practice of engineers. Understanding the "go" involved solving deep secrets of the natural world. As McCulloch told his interviewer:

If you're going to find these things out you've got to get in there and you've got to make measurements, so you've got to have, at your disposal, the physical tools necessary. ... An engineer is a man who has to deliver the goods. And that's what I wanted my instruments to do. So I could learn how to make my measurements.[9]

McCulloch had called upon Maxwell in the past. Besides being the first to "compute the go of the governor of a steam engine," as a scientist, Maxwell was to be emulated. On the potential for using science to solve the mystery of the mind, McCulloch observed:

Even Clerk Maxwell, who wanted nothing more than to know the relation between thoughts and the molecular motions of the brain, cut short his query with the memorable phrase, "but does not the way to it lie through the very den of the metaphysician, strewn with the bones of former explorers and abhorred by every man of science?" Let us answer the first half of his question "Yes," the second half "No," and then proceed serenely.[10]

While McCulloch's views on the value of hypotheses was akin to Maxwell's,[11] McCulloch was more hopeful than Maxwell: metaphysics was not needed, only science. Thus McCulloch's identity as an engineer, and his attempts to understand the "go" of the brain, encompassed both the hands-on experimental mode of practice as well as hypothetical modeling.

The moral economy of McCulloch's collaborative practices also shifted on his move to the RLE, and he continued to act as a mentor.[12] Although he ultimately had mentoring and administrative duties at Illinois, and did collaborative work, he acquired another mentoring identity during this period: that of a sage-collaborator. Collectively, McCulloch's student-collaborators paint a portrait of a man of enormous generosity— intellectually, professionally, and socially. His home in Cambridge as well as the farm at Old Lyme continued to have an "open door policy," and, as Manuel Blum recalls, "a constant flux of hugely interesting visitors."[13] As we saw in chapter 4, McCulloch's style of mentoring was not authoritarian but rather egalitarian.[14] This defined the research culture in his branch of the RLE: he collaborated with students in new ways and attracted a diverse, multidisciplinary crowd of associates with more ties to mathematics, computer science, and engineering than to neurophysiology or medicine.

The fluidity and openness that had always really defined McCulloch, but which never aligned properly with his institutional milieu, finally came to the fore. Neither strictly an experimentalist nor theorist, neither solely mentor nor collaborator, McCulloch's scientific anarchism resonated with the changing disciplinary relations in discussions of the brain and mind that defined the late 1950s and 1960s. However, while some have connected McCulloch's practices to the newly emerging cognitive sciences,[15] in practices that defined the new field of artificial intelligence,[16] and in the new interdisciplinary field of neuroscience—all having important connections to the Cambridge context—McCulloch's relations with these groups was not straightforward.

Finally, McCulloch began to emerge as a public figure, particularly with the publication of a collection of poems, *The Natural Fit*, in 1959, and in 1965 of *Embodiments of Mind*, a collection of McCulloch's essays and poetry. Both were exercises in self-fashioning. From these collections, McCulloch's self-presentation, and the recollections of his peers, McCulloch emerged as an open, courageous, inspiring figure; a polymath who, despite charges of being oblique, controversial, and inconclusive—indeed downright wrong in many senses—spurred a new generation of scientists to go beyond empirics and ask grand humanistic questions about the brain and mind.

## 7.2    A New Institutional Milieu: The Research Laboratory of Electronics

In what kind of research culture did McCulloch find himself at the RLE? To make sense of this we need to step back and look at the culture emerging at MIT. Universities following World War II were different entities, transformed by the new "federal research economy."[17] With the strengthening of pre-existing relations between science, the military, and the federal government, science acquired a new cultural status and new patrons. The political context of the Cold War period saw close collaborations between scientists and the government, and increased federal support for both basic and military research; particularly for projects that were seen to boost national security.[18] This was the era, historians of physics and engineering have traditionally argued, of "Big Science," "Big Weaponry," and the emergence of the military-industrial-academic complex.

Recent scholarship, which pulls back from the traditional focus on physics to include Cold War human, social, and biomedical sciences, reveals a more complicated picture, in two distinct ways.[19] First, historians have rightly begun to question the standard discontinuity story that has defined accounts of pre- and postwar American science.[20] In terms of scale and content, pre- and postwar American science had more in common than has traditionally been argued: "big biology"—biomedical research characterized by large-scale collaboration and alliances with industry—had existed prior to the war, and molecular biology, for example, had existed since the 1930s. Second, historians have begun to move away from an over-emphasis on the Cold War political context for the development of American scholarship (including the natural, human, and social sciences as well as technology). As David Engerman has observed, "the Cold War's impact on the academy was multidimensional, complex, and hardly limited to repression."[21]

On this new view, the Cold War political context both constrained and facilitated the development of science and technology. The federal government directed massive amounts of money, in many instances over the long term, to universities in general, as the baby boom and the GI Bill of Rights, designed to provide educational and financial opportunities for veterans, meant more students and university expansion throughout the 1950s and 1960s.[22] This fueled the unbridled confidence and optimism of many scientists.[23] A key development of Cold War science was the new multidisciplinary and interdisciplinary research formations and research centers.[24] These formations involved exchanges across disciplinary divides that separated the natural, human, and social sciences—particularly the physical and biological sciences—and in many cases these centers were organized and sponsored by the federal government. A related result of this new scientific culture was the fact that often government funded research centers provided a measure of academic freedom for their members.[25] Inter- and multidisciplinary work was an ideal—and as long as the work fit this picture, scholars had freedom to pursue whatever projects they wished, within reason.

At MIT, in January 1946, the new Research Laboratory of Electronics (RLE) was formally established as the wartime Rad Lab closed, and it exemplified the above developments. Besides continuing the model of the Rad Lab, the intention behind the creation of the RLE was to foster a new disciplinary structure for scientific research and move away from specialization.[26] Discussions about the fate of the Rad Lab had been taking place between MIT president and physicist Karl T. Compton, the head of Electrical Engineering Harold L. Hazen, and Physics head John C. Slater, long before the war had even ended.[27] In re-envisioning the Rad Lab, administrators were keen to continue what was perceived as a very successful partnership between two units that symbolized MIT: physics and engineering.[28] Funding for the RLE was eventually secured from the Department of Defense's newly created Joint Services Electronics Program (JSEP), which included funds from the Army Signal Corps, the Navy, and the Army Air Force. The JSEP gave the RLE its first contract of $600,000 per year, for two years, with the possibility of further support, and helped create eight other research centers across the country all devoted to basic research in electronics.[29] The advisory committee of the JSEP understood "electronics" very broadly, and topics covered by the RLE were varied, all falling within the fields of physics and engineering; mostly dealing with microwave physics and high energy physics. Located in the A-wing of Building 20, the structure of the RLE was interdepartmental and interdisciplinary. In the words of its first director, MIT physicist Julius A. Stratton, the RLE vision had stemmed from the recognition that "newly emerging fields of science commonly cut across conventional disciplinary boundaries. And they afford a common meeting ground for science and engineering, for the pure and applied aspects of basic research, to the advantage of both."[30] In its first five years of existence, work at the RLE dealt with microwave electronics, microwave physics, modern electronic techniques, microwave communication, and electronic aids to communication.[31]

By the end of the war, MIT had been steeped in a military atmosphere, largely a result of the relationship that had developed between former MIT engineer Vannevar Bush—architect of American postwar science policy—and Karl T. Compton.[32] Eventually, members of the RLE regularly engaged in applied engineering projects for the US military, such as missile guidance systems, submarine detection, and nuclear powered flight, particularly after the Soviets exploded their own atomic bomb in 1949 and after the mobilization of troops for the Korean War in 1950.[33] Indeed the lines between basic and applied research were increasingly blurred during the postwar period, as sponsored research became "business-as-usual" for scientists at MIT.[34] After a time, however, the RLE was asked by the Department of Defense to expand its scope—eventually these more military "applied" activities were shifted to the new Lincoln Laboratory in 1951.[35]

In reality, however, the picture of research and its relation to military problems at the RLE was less straightforward, and as Roger Geiger has noted, it was more than a continuation of wartime work. Initial plans for the RLE stated that research at the lab was to be basic, unclassified, open, and emphasis was placed on graduate education. Stratton recalled in 1966 that the administrators' vision was "based on a sincere conviction that … basic research was the best possible investment for national defense. … All work under the basic task was unclassified. There was freedom of access, complete freedom of publication."[36] To the question "Why the military?" his answer was, simply, "there was no other support."

The RLE was created at a time when the main federal supporter of science and technology broadly speaking was the Office of Naval Research (ONR). The ONR did more in the postwar period than any other group to support natural science.[37] Although originally established to support development of nuclear power for ship propulsion, during the postwar period, the ONR financed projects ranging from physics to communications, chemistry, physiology, bacteriology, biophysics, and psychology. While the ONR was only authorized to provide contracts, not grants, staff blurred "the operational distinction between contracts and grants by making the Navy's research contracts as unburdensome to the researcher as any grant and as financially rewarding to the researcher's university as any contract a government agency could award."[38] The ONR was a branch of the Navy, and thus within the administrative sphere of the Department of Defense, yet in its early years ONR contracts, as Richard Lewontin has recalled, "were, in fact grants to individual investigators or small groups to carry out research projects generated by intellectual forces internal to the disciplines, provided only that some general relevance to the mission of the federal agency could be established."[39]

The ONR enjoyed a certain level of autonomy from military interests until the rise of the National Science Foundation (NSF) in 1950 and the rise of the National Institutes of Health (NIH), when it began to shift more toward applied research in the interest of the Navy.[40] Further, while during the late 1940s the ONR operated relatively autonomously from the Navy, by 1950, especially just prior to the start of the Korean War, there was more pressure for applied work or research geared toward national security and military concerns.[41] Thus, if we should understand McCulloch's new institutional home at the RLE as being open, "openness" in this context does not simply mean freedom from security clearances and their associated restrictions.[42] Rather, the kind of academic freedom vis-à-vis the RLE and ONR (at least until roughly 1950) meant freedom from "expected public benefits."[43]

By 1951, the RLE had undergone enormous expansion, and the original five research branches of the unit had doubled to ten. According to Wiesner, the appearance of communication research as a distinct area owed much to the work of Norbert Wiener in cybernetics, who "fired up" the communications sciences branch of the RLE, and had for a time visited Building 20 on a regular basis.[44] In 1947, there were a dozen or so RLE members devoted to communication sciences, and summarizing their work in the RLE Quarterly Progress Reports only took a few pages. By 1953, their contributions amounted to 70 pages and by 1963, 230 pages.[45] Claude Shannon joined MIT and the RLE in 1956 and provided the information theory branch of the communication sciences. By this time, communication research had also splintered to include speech analysis, stroboscopic research, network synthesis, processing and transmission of information, mechanical translation, sensory replacement, communications biophysics, transistor circuits, and neurophysiology. Eventually, neurophysiologists, biologists, linguists, economists, social scientists, and psychologists also joined the RLE, all re-conceptualizing their work in terms of communication and information. By 1961, Communication Sciences and Engineering was its own distinct branch of the RLE, and included eighteen research groups (dwarfing the General Physics and Plasma Dynamics groups), including Artificial Intelligence.[46] Five years later, this branch was still holding its own, with Cognitive Information Processing a new group.[47]

## 7.3 The Brain Acquires a New Identity: McCulloch's Transdisciplinary Practices

Almost immediately after his arrival at MIT, McCulloch became less interested in hands-on experimentation, and more interested in strictly theoretical pursuits. Jerry Lettvin claimed that McCulloch's retreat from laboratory work was so dramatic that after 1952, McCulloch never touched physiological experiment again.[48] This flies in the face of McCulloch's self-fashioning as an engineer concerned with measurement and instrumentation yet in this context, the identity of engineer incorporated both theoretical and empirical practices. Although some of his early work up until the end of the 1950s still fell within the realm of experimental neurophysiology, McCulloch's main focus was on the theoretical possibilities of conceiving the brain as a system of neural networks, and for the first time, modeling the brain became the central practice in McCulloch's scientific life. Although McCulloch would characterize his lifelong scientific goal as understanding the functional organization of the central nervous system,[49] at MIT this pursuit was translated into problems of perception, cognition, and reliability. A quick look at the list of McCulloch's publications following his arrival at MIT—of which there are over one hundred with McCulloch as author or co-author—reveals that the majority of his time was spent engaging with the brain from a theoretical or philosophical perspective. Within this new institutional milieu, and with a new collection of scientific collaborators, McCulloch's modeling practices began to further transcend the divides between neurophysiology, communications theory, and engineering.

Both the physical space of McCulloch's corner of the RLE and the disciplinary make-up of the group reflected his transdisciplinary practices. The group that coalesced around McCulloch at the RLE was not a "research school" in the traditional sense. Although this was officially the

"Neurophysiology Group," the space the group occupied at the RLE was not by any means a typical neurophysiology laboratory, nor did the group have any sort of traditional laboratory culture. By April 1958, McCulloch's group had moved into Building 26. As Manuel Blum recalled, McCulloch's room was "filled with books and a blackboard and many desks. And he would have a desk there for his secretary, and his own desk, and he had students like myself at other desks. ... If you went through the door, you'd be in a sort of more closet-like lab which had the frogs in it, and then you kept on going through the next door, and there'd be another lab, also quite large, with this huge cage—copper cage—that was used to keep the radio waves out which you need when you're monitoring a cat's brain. That's where Pat [Wall] would do his experiments." Peter Greene recalls McCulloch's basement office as "dingy" and "cement-floored," containing "four desks pushed together in the middle and one at the side, chairs for these, an old gray sofa, an Oxford English Dictionary in thirteen large volumes, and a high stool on a platform beside a blackboard."[50]

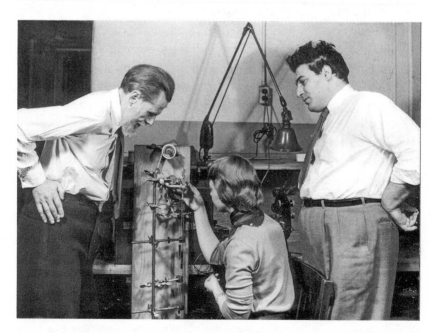

**Figure 7.1**
McCulloch at the RLE with Jerome Lettvin and an unidentified woman (n.d.). MIT Museum.

In this multidisciplinary environment, McCulloch's scientific practices were diverse, and fall into three related areas that span disciplinary divides: experimental work on visual and auditory systems (mainly on cats and frogs), theoretical modeling of the central nervous system (including synaptic transmission, the reticular formation, and reliability), and rhetorical work pushing the cybernetic creed and its relevance for understanding the brain and mind. His continued practices of self-fashioning also form part of this set of activities. Taken together, these latter practices involve McCulloch taking on new roles as a transdisciplinary scientist spanning the divides between science, medicine, engineering, and philosophy. While McCulloch had engaged in the pursuit of models in the past, and he was still pursuing questions about the functional organization of the central nervous system, what had changed with his position at the RLE was an institutional space where transdisciplinary work was encouraged and supported. Now, McCulloch made sense of the functional organization of the nervous system in terms of circuit theory.[51]

### 7.3.1 Modeling Neurophysiology

The nucleus of McCulloch's group during the 1950s had Pitts doing theoretical and mathematical work and Lettvin designing the electrical apparatus for recoding and amplifying signals and experimenting on single cells. By 1954, Wall was also working with afferent neurons, reflexes, and transmission of impulses.[52] Humberto Maturana joined the group during the late 1950s. Most publications based on experimental work were co-authored with Lettvin or Pitts, and judging from Lettvin's recollections, McCulloch's name appeared here likely out of respect.[53] This work occupied McCulloch very soon after his arrival at MIT, through to 1960. Of course, the most famous paper arising out of this work was the 1959 work, "What the Frog's Eye Tells the Frog's Brain." The paper demonstrated the existence of "feature detectors" in the optic nerve of the frog, specific neurons that respond to specific features in a visual stimulus—movement, contrast, curvature, and dimming. Although the author list for the paper included McCulloch and Pitts, the paper was actually the product of a collaboration between Lettvin and Maturana. Maturana and Lettvin spent a year developing a way of recording from single unmyelinated optic nerve fibers in the frog. By 1957, they finished, and along with input from McCulloch, Pitts, and Oliver Selfridge, they were ready to publish. This work was highly empirical and required a good deal of experimental drudgery.

Lettvin's work with Maturana was actually a departure from the original logical approach that McCulloch and Pitts had taken to the brain and perception. As Lettvin recalled, instead of a model of the mind that involved a system of logical elements, his work with Maturana suggested that in perception, "the world was reported by natural language rather than logical language ... in terms of things and their relations rather than sense data to be processed into the concepts of things."[54] That they discovered that a "fixed five-layer system of elements, laterally interacting, gave the content of perception to a frog" suggested to Lettvin and Maturana that they were dealing with an analog, not logical, system.

### 7.3.2   Reliability

Another concern for McCulloch at the RLE was to make sense of the brain's organization in terms of stability in biological systems. This facet of his modeling practices involved the development of probabilistic logic and circuits that are logically stable under shifts of threshold. Manuel Blum, who spent several years working as an assistant in McCulloch's group at the RLE, says that McCulloch was

trying to design neuronal circuits that would function correctly despite errors. ... He said, "When you drink coffee your thresholds go down; when you drink alcohol your thresholds all go up. And yet you can still walk, you can still talk, and you can still think."[55]

A vivid and apt analogy for McCulloch—he was reported to live on whiskey and ice cream—in fact in this work McCulloch was translating the problem of the brain's organization in terms of the stability, the reliability, and the flexibility of nervous activity.[56] Reliability had been a key question in Cold War engineering since the early 1950s, and emerged as a problem in both human and technological spheres.[57] Shortly following the start of the Korean War, a broad fallibility was perceived in defense technology and the electronic infrastructure of the Cold War, generating what contemporaries saw as a "reliability crisis."[58] Thus, steeped in the atmosphere of the RLE, it is not surprising that McCulloch would translate a biological problem into an engineering one, and vice-versa.

Besides Blum, aiding him in this work was Scottish graduate student Jack Cowan and a Dutch graduate engineer named Leo Verbeek. Within this framework of circuit theory, the brain became a device for transmitting information. Using probabilistic logic—a departure from his earlier logical models—McCulloch developed models in which neural connections remain stable despite shifts in threshold and those which display more reliability than their components.[59] McCulloch had already departed from the straightforward Boolean logic used in his 1943 paper, even in his 1947 model of perception with Pitts, in which the processes were less logical and more statistical—less about single units and pure logic—and more about collections of units and patterns of activity. As Margaret Boden observes, the 1947 model did not deal with how we *learn* universals, only with how we *recognize* them.

McCulloch's later search for a probabilistic logic in part came from the failure of the 1947 model. In the mid-1950s and just before his death in 1957, John von Neumann had begun applying probabilistic logic to networks of neurons. Here he argued that brains and digital computers are sufficiently different that our models of the brain cannot be based on pure logic.[60] Complicating this picture, for von Neuman, are the facts that there is an enormously high number of neurons in the CNS, neural thresholds are likely not constant, neural connections are multiple and complex, the synapse between two neural elements is more analog than digital, and the nervous message is statistical in nature.[61] Probabilistic logic had the potential to deal with these issues. As McCulloch argued in 1961, in probabilistic logic, not merely the value of the argument, but the function itself is more or less probable. Whereas the 1943 logic had the relation between two units fixed—for example, the logical function "A and B" is constant—to more accurately model the brain, we need to have the relation between two units probable, and not all possible functions are equally probable.[62] In this way probabilistic logic can account for shifts in thresholds, the stability, the flexibility, and even the infallibility of the CNS.

Again, McCulloch continued his practice of hypothetical modeling in his work on this topic, a practice he explicitly likened to Galileo's initiation of "modern physics in terms of hypothetical interactions of postulated entities."[63] His research plan, although falling within the realm of neurophysiology, was also a project in engineering and mathematics and took advantage of the multi-disciplinary climate at the RLE. In a communication to the officers of the National Institute of Health, from which he'd received some funding for his experimental work on nerve transmission and vision, McCulloch expressed his satisfaction at his new environment:

> I am where I want to be, for I have daily contact with the best theorists in the fields of organization—circuitry, coding, information, games, automata, computers, and servosystems. I have had the best team play with experimentalists of my own selection. We are superbly equipped to make precise quantitative measurements to discover the mechanisms of the interaction of neurons, their intrinsic properties and those of their parts.[64]

Although the subject of investigation was the brain, the theories themselves, as McCulloch presented them to the National Science Foundation, are "logical structures concerned with questions of organization in the field of communication by means of signals."

### 7.3.3   The Reticular Formation

In 1949, neurologist Giuseppe Moruzzi and neurophysiologist Horace Magoun of the Northwestern University Medical School had used localization and stimulation techniques to demonstrate that electrical stimulation of a brain stem system of neurons—known as the reticular formation (RF)—produced EEG patterns nearly identical to those produced during transitions from sleep to wakefulness.[65] Many brain researchers concluded that the reticular formation had something to do with consciousness and was thus a fundamental organizational structure in the brain. A flurry of scientific work followed on the heels of their work. As one contemporary recalled, "[t]he problem of the relationship between the brain and consciousness appeared in a new light. Everybody rushed to find out about the reticular system."[66]

However, by the end of the 1950s, the promise of the reticular formation for explaining mental experience had faded. Nevertheless, at the RLE, McCulloch took up study of the RF with Bill Kilmer, translating it into the cybernetic language of controlled loops of neural units, via positive and negative feedback.[67] This model presented the RF as responsible not for consciousness per se, but for behavior choice. McCulloch and Kilmer, eventually working with Blum, designed a hypothetical reticular formation as a model decision system that issued "signals" to govern behavior. Based on neurophysiological and anatomical knowledge of the reticular formation, and building on an idea of Roberto Moreno-Diaz's, their model treated neurons as "nonlinear oscillators" and set up a computer simulation to model four modes of behavior.[68]

During the late 1950s and 1960s, McCulloch also did work as an advisor and lecturer for the US Air Force Avionics Laboratory at the Wright-Patterson Air Force Base in Ohio—mainly on bionics. He received support in the form of a contract from the US Air Force Office of Scientific Research, specifically the "Advisory Committee for the Directorate of Information Sciences" beginning in 1958. In the summer of 1960 this enabled him to attend several conferences in Europe and Britain, spend time with British cyberneticians Gordon Pask and Stafford Beer, and report on their activities.[69] He also lectured at several Veterans Administration Hospitals, work he had continued from his time at the Illinois Neuropsychiatric Institute, where he'd been enlisted by the US Army Medical Division (the Office of the Surgeon General) to present to students at the Medical Service Graduate School in Washington, DC, on the circuit action of the nervous system and the basic sciences fundamental to psychiatry.[70]

Within McCulloch's new modeling practices and collaborations—on vision, synaptic transmission, reliability, and the reticular formation—the brain as a scientific object acquired a new identity. As McCulloch's institutional milieu was steeped in multidisciplinarity and framed by a union between physics and engineering, the language he used to engage with the brain was that of communications, information theory, and electronics.

## 7.4   McCulloch and the Cognitive Sciences: A Complex Legacy

Given this focus on hypothetical models of the mind, it is tempting to present McCulloch and his work as "foundational" in the emerging fields of artificial intelligence and the cognitive sciences of the late 1950s and 1960s.[71] McCulloch's practices and transdisciplinary perspective resonated with this work, yet was distinct in both spirit and content, in important ways. Admittedly, two key commitments of cognitive science—the focus on "complex internal cognitive processes" such as perception and memory, and the commitment to the metaphor of computation[72]—are central features of McCulloch's explanatory framework. Although scholars have rightly questioned the applicability of the notion of a "revolution" in describing the history of the human sciences during this period, historians generally agree that a stark discontinuity existed between neo-behaviorist psychology of the 1930s and 1940s and the "cognitive" psychology that began to emerge in the second half of the twentieth century, with the appearance of two oft-cited "landmark" studies: *A Study of Thinking* (1956) by Jerome R. Bruner, J. J. Goodnow, and G. A. Austin, and *Plans and the Structure of Behavior* (1960) by George A. Miller, E. Galanter, and Karl H. Pribram.[73] Indeed, early "cognitivists" such as psychologists Bruner and Miller redefined mind in terms of information processing.[74]

Certainly, McCulloch was also after such re-conceptualizations of mind and cognitive functions. Further, Bruner and Miller also had a bias toward interdisciplinary practices—Miller had worked in cross-disciplinary research units such as MIT's Lincoln Lab, which, like the RLE, sprung from a Cold War research culture. Bruner and Miller had had interdisciplinary motivations in creating the Center for Cognitive Studies at Harvard. Like McCulloch, Miller had been influenced by information theory, and Bruner had been influenced by von Neumann's game theory.[75] Yet while McCulloch was also interested in mental functioning, he was fundamentally a brain researcher. In contrast, Bruner's and Miller's practices were devoted to understanding *thinking* and *learning* first and foremost. They were driven by a need to reform psychology, to address the neglect of the mind by behaviorists and import the study of cognition back into psychology.

In principle, McCulloch's vision for understanding the mind through logic, modeling, mechanical analogies, and information processing also resonated with the goals of early figures in AI, such as Marvin Minsky (who worked with McCulloch at the RLE), Herbert Simon, and Allen Newell.[76] However, McCulloch was always fundamentally interested in the brain, unlike the early AI or cognitive science pioneers. McCulloch certainly had an influence on some of these figures: his early work with Pitts, for example, was often referred to as one of the first explicit outlines of the notion of artificial intelligence. In the August 1955 proposal to the Rockefeller Foundation for a Summer Project at Dartmouth College on Artificial Intelligence, authored by Claude Shannon of Bell Labs, mathematician John McCarthy, Marvin Minsky (then a Harvard Junior Fellow in mathematics), and N. Rochester of IBM, McCulloch was included as a potential participant, although he was not at the meeting.[77] This outline defined the "artificial intelligence problem" in terms of programming automatic computers to simulate machines, programming computers to use language, arranging neuron nets to form concepts, developing criteria for efficiency of calculation, the question of self-improvement, abstractions, randomness, and creativity. Yet the connection to the McCulloch–Pitts work was not straightforward. For example, McCarthy claims that one of the reasons for using the term "artificial intelligence" to encompass these problems was to actually *escape* the association with cybernetics: "[i]ts concentration on analog feedback seemed misguided, and I wished to avoid having either to accept Norbert Wiener as a guru or having to argue with him."[78]

Early on, boundaries between AI and cybernetics were blurry, and since AI's status as a scientific field still open to question, some figures in the nascent AI group attempted to delineate their aims from those of the cyberneticians.[79] In a perceptive analysis of the shifting relations between cybernetics and AI, Ronald Kline has examined the history of the Dartmouth Project and highlights the fact that during these early years (the mid-1950s), participants in the Dartmouth conference—often seen as the "birthplace" of AI—struggled to delineate the boundaries of their field and distinguish their practices from those of cybernetics, which, in the end, was characterized by "brain modeling." In the picture Kline presents, as a field, early AI emerged as chaotic and far from uniform or coherent. At Dartmouth, Shannon, Minsky, and Rochester still showed interest in brain modeling while McCarthy, Newell, and Simon were already leaning toward symbolic processing—even in the proposals, as Paul Edwards points out, the "still-nascent split between the computer-brain and computer-mind metaphors already appears clearly."[80] A rift was also apparent a few years later, in November 1958, at a meeting on "Mechanisation of Thought Processes" at the National Physics Laboratory at Teddington, Middlesex, England, where both Minsky and McCulloch presented.[81] Many AI pioneers attended, and McCulloch presented his work on reliability yet was attacked by the new crop of AI researchers. Minsky had abandoned brain modeling by 1958.

By the early 1960s, the growing field of AI was beginning to dominate discussions of computers and brains. In 1963, for example, the editors of an anthology of papers on *Computers and Thought* drew an opposition between the pursuit of "cognitive models" and the development of "neural cybernetics" or "self-organizing systems," all falling within the category of artificial intelligence.[82] Those who build cognitive models, according to the authors, take a "macroscopic" approach to the development of intelligent machines, "using highly complex information processing mechanisms as the basis of their designs. They believe that intelligent performance by a machine is an end difficult enough to achieve without 'starting from scratch'," whereas neural cybernetics focuses on much more simple information processing elements arranged in random networks. The edited volume included papers by Noam Chomsky, Minsky, Newell, Oliver Selfridge, and Herbert Simon, and few in the neural modeling field. Minsky compiled the bibliography for the book, and while the macroscopic approach dominated the volume, Minsky's bibliography included over twenty of McCulloch's papers.

Minsky had spent at least a year in the midst of McCulloch at the RLE. McCulloch's scientific legacy with AI is complex and nuanced, as Minsky, twenty-nine years McCulloch's junior, perceived and received from McCulloch an appreciation for working outside of existing disciplinary boundaries. He recalled traveling with "this wonderful man" who gave him "a new view of the world."[83] Minsky's original interest in neural networks had come from reading McCulloch and Pitts as well as Rashevsky. Once associated with him, Minsky drew much more, scientifically, from McCulloch. As Minsky recalled, one day McCulloch told him:

I'm going to explain something to the psychologists. And he had a sort of grand view of this, the importance of cybernetics, which was correct, so, otherwise you would have said he was delusional. And so I went with him, and it was a little meeting with about six or seven people ... but he was on the stage, and I realized he was talking to the whole world. Normally people are talking to their audience but he would have very elaborate constructions and beautiful ways of saying things. ... I've never met anyone for whom all the world was a stage. ... From that I got some sense that you shouldn't waste people's time with things that aren't very important ... of course nobody could live out such a commandment, but I must have spent most of a year just hanging around him and trying to understand how he could see such importance in ordinary things.[84]

Practically speaking, McCulloch was a brain researcher, yet constantly strove to connect his inquiry to grander questions—to make the study of the brain of existential, humanistic consequence. He practiced openly and hopefully. He was not arrogant—but imbued his scientific work—and the pursuit of science—with cosmic significance. This facet of McCulloch's identity says much about his ability to attract both disciples and detractors.

## 7.5   A Sage-Collaborator

One of the most important facets of McCulloch's legacy does not lie in a burgeoning field of cybernetics, or even in one definitive grand theory or solution to the problem of mind and brain. Rather, it lies with a diverse group of practitioners whose work has been shaped by McCulloch's practices of hypothetical modeling—either through direct mentorship or indirect influence. Toward the end of his interview with Percy Saltzman on the CBC program *The Day It Is* in 1969, McCulloch was asked for his thoughts on the legacy of his work. Smiling, he replied "Oh, that's very easy. It's the youngsters who work with me. I've had a hundred and twenty younger— more than a hundred and twenty—younger collaborators. Forty of these are very good scientists in their own right today." Indeed, McCulloch mentored a new "crop" of youngsters who eventually became pioneers in studying the brain from the perspective of computational complexity.[85]

By the time of McCulloch's arrival at MIT, children Taffy, David, and George had gone to college, while Mary Jean stayed on in Old Lyme, where many local youth benefited from workshops on the farm.[86] With a nearly empty house, McCulloch adopted more student collaborators. Eventually while at the RLE, besides his long-standing collaborators Walter Pitts and Jerry Lettvin, a diverse group of student-collaborators began to coalesce around McCulloch.[87] McCulloch had always been actively engaged in supporting students, and this practice was arguably a facet of his motivation for coming to MIT. McCulloch used much of the energy and funding he received while at the RLE to encourage and support his young colleagues. For example, many of the funds McCulloch received from the ONR during the early 1960s were used for travel expenses to "promote and aid research in the field of Neurophysiological research."[88] A non-exhaustive list would include, of course, Lettvin and Pitts, but also Pat Wall, Manuel Blum,[89] Michael Arbib,[90] Jack Cowan,[91] Roberto Moreno-Diaz, Marvin Minsky, Oliver Selfridge, Seymour Papert, William Kilmer, Francisco Varela, Leo Verbeek, Donald MacKay, Stuart Kauffman, and Humberto Maturana. While these were all junior scientists, with training in fields as diverse as mathematics, neurophysiology, engineering, and computer science, McCulloch treated them as equals and encouraged their intellectual growth: as Blum recalled, "[one] thing that made Warren truly unusual, something that only few professors do at all, let alone well, for which I am truly grateful: he raised up quite ordinary people (like me) and made them feel special."[92] McCulloch was also an inspiration to numerous other figures in a number of disciplines related to studying the brain and mind during this period and his influence went beyond the walls of the RLE. As McCulloch engaged in this interdisciplinary activity in his collaborations, he began to instill in his students an appreciation for this style of scientific practice. Many of the students drawn to McCulloch's group at the RLE came from a mathematical or engineering background. Although Blum had been doing his Ph.D. work in mathematics at MIT with Minsky, McCulloch made more of a lasting impression on his scientific work and beyond:

Marvin was very important to me, but Warren was the most important for a number of reasons. But probably the most important was he really treated me like an equal. I mean I was by no means anywhere near being his equal but he treated me like that. He actually listened to what I had to say. He made great suggestions. And I could actually read his stuff, I mean after all I was just a junior, and at that time mathematics was really hard, I could hardly read it, but I could read Warren's stuff.[93]

McCulloch, according to Blum, "looked and talked like God," yet "treated everyone he touched with a dignity far beyond our due."[94] McCulloch's open, liberal approach to parenting colored his role as a mentor—his children called their parents "Warren" and "Rook," and he encouraged their self-reliance and responsibility as members of the family. As an egalitarian, "laissez-faire" mentor—a role that says as much about his style of mentoring as his role as a scientist—McCulloch similarly elevated his students' roles as not junior scientists but collaborators. As Laura Otis has shown in her study of Johannes Müller and his students, descriptions of scientists that are based on students' recollections cannot be taken as passive descriptions but as active interpretations that reflect the interests of the student and their understanding of themselves as scientists.[95] These recollections, then, are not straightforward descriptions but rather a carrying on of McCulloch's scientific and intellectual tradition.

## 7.6   A Polymath

While at the RLE, McCulloch was still involved with the medical and psychiatric community and even with the psychosomatic research community.[96] However, it was soon clear that he was leaving this world behind.

Still in contact with psychiatrist Eugene J. Alexander of the Henry Ford Hospital in Detroit, who invited McCulloch to speak at the March 1954 meeting of the Michigan Society of Neurology and Psychiatry, McCulloch wrote to him about the subject of his talk, which would stress the "Physical Treatment of Mental Diseases," which, McCulloch lamented, physicians do not recognize properly as the best way of dealing with psychotic disorders.[97] McCulloch suggested he might give a talk on carbohydrate metabolism and its role in diseases of the brain, and reflected that "it would be most fun to do something entirely different and tackle either as a problem in circuit design the construction of ethical robots or something equally frivolous normally supposed to be purely mental but, to say the least, manageable by purely physical means. This often does manage to bridge the chasm between mind and body. ..."[98]

Alexander responded enthusiastically to McCulloch's ideas. Like McCulloch, he was part of the anti-Freudian faction in American psychiatry, and indeed had an interest in the problem of "how the brain mechanically performs the functions we call mental." Further, he wrote,

I believe any modern psychiatrist would accept readily the concept that the brain is the organ of the mind, and that the mind is simply the functioning of the brain. However, I also believe that most psychiatrists would consider it "frivolous" to attempt to make a working model of the brain. The mind is still somehow sacred as the soul once was. ...[99]

Alexander warned McCulloch that the title "Towards Some Circuitry of Ethical Robots" would "frighten our membership." Presenting, perhaps, the same ideas, but within the framework of carbohydrate metabolism would be better, and presenting it in even more a clinical framework would be even better still. Essentially Alexander wanted McCulloch's talk to "kill two birds with one stone"—appealing to doctors treating patients, something interesting and "applicable to their work" but also something for one who knows little neurophysiology and even less mathematics, something on how the brain "accomplishes the functions of the mind."

Such addresses comprised an important and regular facet of McCulloch's scientific practices and identity during this period: acting as a spokesman for cybernetics and a promoter of his particular style of modeling. Much of this involved a level of self-fashioning. For example, during this period we see McCulloch beginning to weave his own autobiographical trajectory with the biographical story of cybernetics.[100] This practice is most striking in the appearance, in 1965, of a collection of twenty-one of McCulloch's papers, *Embodiments of Mind*. McCulloch's visage on the cover of the book captures his persona at this stage of his life—a high-contrast, black-and-white photo of a white-bearded man, with creased brow and sunken cheeks and eyes, staring at the viewer with a piercing gaze. Most of the papers are public addresses on the functional organization of the central nervous system, and cast his life's work as a search for the solution to the mind-brain problem.

*Embodiments of Mind* was an exercise in self-fashioning—generating, for McCulloch, the image of a polymath. McCulloch promoted the cybernetic style of scientific practice continually cast in the language of philosophy and allusions to literary themes and references, with address titles such as, e.g., "Information in the Head," "Through the Den of the Metaphysician" (a gesture to Maxwell), "Towards Some Circuitry of Ethical Robots," "Mysterium Iniquitatis of Sinful Man Aspiring into the Place of God" (a gesture to Catholic doctrine), "Where Is Fancy Bred" (Shakespeare),[101] "Agathe Tyche: Of Nervous Nets—the Lucky Reckoners" (Greek mythology). By now, with no professional obligations to psychiatry, such addresses were apparent displays of virtuosity—shifting continually between the language of cybernetics, engineering, philosophy, neurophysiology, and psychiatry. Many of these talks are dense and barely penetrable. An address from 1961, to a group of psychiatrists, concerned, what he told his audience, "that logic which we need in psychiatry."[102] Its title, "Abracadabra," referred to "the use of symbols to affect man's conduct." For McCulloch, the circuit theory of brains and its accompanying logic held enormous promise for a rigourous, respectable framework for psychiatry. With such titles (and requests for public lectures), McCulloch was performing the role of a public figure, a spokesperson for the cybernetic creed and a guru of theoretical, grand questions about the brain, its organization, and what it is to be human—indeed, he was speaking to the entire world.

Included in *Embodiments of Mind* was a collection of McCulloch's poetry. In December 1957, McCulloch read a selection of his poems—mostly sonnets—to the Chicago Literary Club. The collection was published as *The Natural Fit* in 1959. The title signifies a union, for McCulloch, of poetic and scientific activites. He is likely using the term fit in its literary sense of a *fytte*, simply a part or section of a poem or song. In his introduction of the book and in later subsections, McCulloch invokes the term "fit" in references to historical figures in literature: the legend of "The Sophomoric Fit" comes from Boetius, *The Consolation of Philosophy* (AD 888); "The Outlandish Fit" from Chaucer; and finally "The Natural Fit" from Samuel Johnson's *Rasselas, or The Prince of Abyssinia* (1759), a meditation on happiness and how to attain it. In Chapter Ten of Johnson's narrative, the prince, confined to Happy Valley, could not find hope or delight in anything, and decided to escape. He meets Imlac, a man of learning, a poet. In telling his story to the prince, Imlac reflects on poetry as the "highest learning."[103] In trying to emulate early poets, Imlac realized that imitation was not enough: he decided that "[n]ature was to be my subject, and men to be my auditors. ... All the appearances of nature I was therefore careful to study. ..." After relating this and a long list of requirements for being a poet, "Imlac now felt the enthusiastic fit, and was proceeding to aggrandize his own profession, when the prince cried out 'Enough! Thou hast convinced me that no human being can ever be a poet. ...'" "To be a poet," said Imlac, "is indeed very difficult." "So difficult," replied the prince, "that I will at present hear no more of it."

In this story, a poet is not only a literary figure but a student of the natural world. Similarly for McCulloch, his performance of the identity of a poet was simply an extension of his role as a scientist, a form of scientific expression. In *The Natural Fit*, McCulloch quoted this last exchange, adding in an autobiographical vein, "The form of the sonnet is difficult, but the required compression of his enthusiasm was natural to his way of thinking hot."[104] Fit here takes on a different meaning, as a sudden or transitory state of activity, a paroxysm.[105] Working in a traditional form, restricted by the rules of the sonnet, and gesturing to grand philosophical questions, McCulloch defied conventions by transcending modes of expression between the scientific and the poetic.

In the introduction to *The Natural Fit*, McCulloch, nearly sixty, alluded to his past as three characters, who, besides having a common interest in "communication," had little else connecting them.[106] This is a performative act, whereby McCulloch the engineer is reinterpreting his past scientific efforts in the language of the communication sciences. The last poem in this collection, likely written in 1957, sees McCulloch reflecting on his position as a senior scientist:

farewell sweet morrows hopes deferred and all
crisp years fat earnest in defect of youth
indian summers quicken to keen fall
as brisk october blazons times no ruth

i cry no quarter of my age and call
on coming wits to prove the truth
of my start venture into fates cold hall
where thoughts at hazard cast the die for sooth

from me great days are gone and after none
array the ardour that i scarce compress
in temperance terrible charged i abide
the desperate victor of my last race run
wanting bold challenge to life's dread excess
to fire that frenzy I must else wise hide[107]

The second stanza captures McCulloch's role as a mentor during this period. By the late 1950s, it was clear that McCulloch saw that he had accomplished scientifically most of what he was to accomplish in his life, and that he was ready to "call on coming wits" to continue what he'd begun and "prove the truth of my start venture into fates cold hall."

McCulloch had sent out his collected poetry to friends at Christmas 1959. They were favorably received with delighted surprise, and friends expressed appreciation for this dimension of McCulloch's scientific and intellectual life. Heinrich Klüver wrote, "I am glad I that I finally met Warren #2. You do not even know how glad it made me! Of course, I suspected the #2 existence but did not know of its glory until I read the poems last night. Thank you!"[108] John F. Fulton, McCulloch's former colleague at Yale, wrote that he did not know if he had "ever been more pleasantly surprised than I have been by the perusal of your poetical autobiography, *The Natural Fit*."[109] Fulton added he wanted to have someone review the collection for the journal he edited, *Journal of Neurophysiology*, and thought that it would add "the kind of humanistic touch which I have always liked to see in the Journal and have been so seldom able to include. ..."

Reviews of *Embodiments of Mind* appeared by fellow cyberneticians W. Ross Ashby and W. Grey Walter, and neuroscientist Karl Pribram, who had worked with Karl Lashley. They were all enthralled with the book. According to Ashby, those who like being "stimulated, goaded, even infuriated into developing new ideas" will enjoy it; those who "like to rest comfortably in the established and obsolete" will not.[110] Some did not—arguing that the titles of McCulloch's essays "promise more than his discourse yields."[111] In a particularly eloquent review, Walter denied the assessment of McCulloch as an eccentric, rather, "he is the antithesis of eccentric, he is at the centre, the pivot of a whirligig of explosive thinking."[112] Walter goes on to ask: "Why should he be modest? Who else in this century of destructive progress has created so much? Who else can joke with algebra among the neurons, ridicule Freud among the analysts, play with prosody among the poets?" Walter deemed McCulloch not the Newton of neurophysiology but its Swift. There is little in "clinical innovation or discovery" in McCulloch's work, yet we are still indebted to him for "more than a pile of cheap facts; he has endowed us with more capital ideas than we can spend the interest from in a generation."[113]

## 7.7 Conclusions

The institutional research culture at the RLE allowed McCulloch's scientific identity to evolve in several ways. Whereas at Illinois, he played the role of biological psychiatrist—and in parallel the cybernetician, with modeling being peripheral to his laboratory work—at MIT, McCulloch's modeling activities defined his scientific identity. Situated at the nexus between communication theory, information theory, neurophysiology, and engineering, McCulloch performed the identity of an engineer, relating the reliability of technological systems to that of the nervous system.

At the RLE, McCulloch had the institutional and professional freedom to pursue various forms of scientific expression. By the late 1960s, McCulloch was a tall, lank, God-like figure, with a long unfashionable white beard, rumpled clothes, usually with a cigarette dangling from his fingers. Beyond his activites as an engineer, now in his sixties, McCulloch performed the roles of sage-collaborator, polymath, and poet. These were performances—simultaneously products of his own agency and his institutional context—both private and public. Institutionally, McCulloch's group had a complex, mixed funding model and diverse patrons with no strict separation between physics and biomedicine; a new kind of alliance between the two that was distinct from the traditional example of molecular biology. Within this complex space, the brain acquired new identities as a scientific object, in much the same way as thinking, behavior, intelligence, and memory did for early artificial intelligence pioneers. This patronage pattern also had significant effects on the sciences of the brain by creating interdisciplinary spaces for new sorts of practices to emerge. Previously dominated by support from philanthropic organizations such as the Rockefeller Foundation and the Macy Foundation, and almost exclusively motivated by social concerns about mental illness in the American population, during the postwar period, brain research increasingly drew support from the federal government. Although he was working in a "laboratory," the picture of McCulloch as an experimenter here does not fit, nor does the picture of McCulloch as a strict theoretician, or even a rhetorician.

This institutional structure facilitated McCulloch's practices as an engineer, but also as a mentor, giving McCulloch and the group that coalesced around him freedom and flexibility, without disciplinary restrictions. Now more of an elder statesman of science, his role in the RLE was more as a mentor, one that was liberal, "laissez-faire," and egalitarian.[114] McCulloch's evolving identity played a role in his style of scientific pedagogy at the RLE. For many of the scientists he collaborated with and mentored, he created an image of the scientist that asked big questions, transcended disciplinary divides, and engaged in scientific practices that meshed the technical skills of laboratory research with the framing of questions that saw a bigger picture—in essence asking questions about the brain that went beyond what was possible in the laboratory and necessitated the use of modeling as a scientific practice. The institutional framework at MIT and the RLE, which fostered interdisciplinarity and fuelled an "anything-goes" attitude, facilitated these changes. McCulloch's relationships with his student-collaborators shows that his legacy as a cybernetician lies as much with his social and intellectual generosity as with his scientific contributions. Manuel Blum has said that while many McCulloch's ideas from this period may not hold up to experimental scrutiny, they were nonetheless "perceptive, exciting, and stimulating … they demonstrated a way of thinking about these things that was crucial for my own development." Neuroscientist Karl Pribram's assessment was similar—McCulloch's scientific exercises were "brilliantly conceived and brilliantly carried out," and often, "brilliantly wrong. But that does not matter."[115]

# 8 Epilogue

## 8.1 McCulloch's Identities: Transdisciplinary Practices

Warren McCulloch performed multiple identities through the course of his scientific life: student, poet, experimental neurologist, psychiatry resident, neurophysiologist, aspiring scientific philosopher, neuropsychiatrist, administrator, cybernetician, modeler, engineer, and mentor. None were simply a product of his own agency, nor were they a straightforward result of his varied institutional contexts. Rather, they were performatively produced.[1] Performative identity is acquired through doing and being: in McCulloch's case, through both his scientific practices and personal activities. He moved through these disciplinary worlds in part because of institutional fluidity, in part because of his own ease at transcending intellectual fields, and his various identities were only stable insofar as he was in a predictable disciplinary or institutional context.

By the 1960s, McCulloch had presented his life as a continual search for a scientific solution to the mind-body problem, cast as a physiological theory of knowledge, an experimental epistemology. In characterizing McCulloch's self-fashioning as a performative act, I am less interested in demonstrating that McCulloch was *not* always interested in the mind-body problem. Rather, I have interpreted McCulloch's practices surrounding this question in terms of disciplinary relations. Many of the identities McCulloch performed were disciplinary identities. Disciplines, to borrow a phrase from Clifford Geertz, are ways of being in the world—whether that discipline be in the natural sciences, medicine, human sciences, or philosophy.[2] To engage in particular scientific practices "is not just to take up a technical task but to take on a cultural frame that defines a great part of one's life." McCulloch's scientific life was transdisciplinary in several ways. First, he posed grand questions that transcended the divides between science, medicine, and philosophy. He continually cast and recast his work as a search for the solution to the problem of brain and mind. Second, he employed hypothetical modeling as his main methodological tool (while at the same time being well versed in laboratory practices). Finally, McCulloch made use of transdisciplinary rhetorical strategies, using deliberately ambiguous language in which terms have multiple meanings, in order to appeal to different disciplinary audiences. While this may have been a deliberate linguistic choice, such strategies also reflect his own ease at moving across disciplinary boundaries—whether it be between clinical and basic research, empirical and theoretical practices, science and philosophy, medicine and information sciences.

Indeed, if one steps back from McCulloch's self-fashioning as a scholar driven by the philosophical problem of mind, we can see that it really expressed sensibility and irreverence about the divisions that organized American academic life. As Roger Smith has observed, "inquiry into the regions where bodily and mental regions overlap ... has also been inquiry into what the relations of human and natural sciences should actually be."[3] When one turns McCulloch's rhetorical gesture to the mind-brain problem inside out, one can see a more overt concern with blurring the boundaries between science, medicine, and philosophy. By being transdisciplinary, McCulloch mounted a critique of the world-view of modern science and thus more specifically essentialism about modern scientific disciplines.[4]

Why cross disciplinary boundaries? One might do so if one shares common interests with members of other disciplines, to overcome "tribalism," or to search for tools when the problem to be solved cannot be dealt with in the realm of one's own discipline.[5] Although McCulloch transgressed disciplinary boundaries repeatedly in his career path, in his rhetorical practices he also oscillated *between* them, shifting between the discourse of science, medicine, and philosophy. Recasting philosophical problems in the language of science can operate not only to reinforce and create boundaries between disciplines but also to permeate them.[6] This, in fact, was at the heart of McCulloch's rhetorical spin on cybernetics. Ian Hacking has articulated two senses in which we can speak of scientific unity—in terms of singleness or harmony. Cyberneticians of McCulloch's ilk were after both: a single style of scientific practice—mathematical modeling—and transdisciplinary harmony.[7] Throughout the cybernetics movement, core members of the group attempted to explain phenomena in their fields in terms of the "circular causal mechanisms" of the engineering sciences. Thus, on a methodological level, the perceived "unity" of cybernetics was directional.

Some of McCulloch's transdisciplinary aspirations were already being realized in the decades following the Second World War. New disciplines, new disciplinary alliances (e.g., between psychology and physiology), and new scientific practices transformed the human sciences and the study of brain and mind, in part a result of the open-ended research culture of the period.[8] The use of the term "revolution" to describe this process has been challenged, for several reasons: first, explanations of behavior that involved mental, internal states were really only a big departure from the norm in psychology, not in other human sciences. Further, the idea of a revolution assumes a departure from a monolithic behaviorist psychology, which did not exist. Secondly, the changes that took place were much more piecemeal than the image of a revolution suggests, and involved the gradual emergence of the cognitive sciences during the 1960s—a multidisciplinary endeavor that includes cognitive psychology, neuroscience, linguistics, computer science, philosophy, and artificial intelligence.[9] Thus in the wider arc of McCulloch's story, we see the quests for disciplinary unity realized in a multiplication of multi- and interdisciplines devoted to making sense of the brain and mind—many operating outside the strictly biomedical context that had dominated studies prior to World War II.

In terms of the sorts of questions being asked, McCulloch's practices were out of step with some of these early endeavors. Early AI researchers—who practiced what some have called GOFAI (Good Old Fashioned Artificial Intelligence)—had more interest in artificial systems, processes of thinking, problem solving, and planning; and less interest in the brain itself.[10] Other neuroscience discipline-builders and visionaries—such as MIT's Francis O. Schmitt—who coined the term "neuroscience" in the early 1960s—had a more firmly integrated and molecular approach, which wasn't part of McCulloch's project at all. So where can we connect McCulloch's practices on the diverse map of cognitive sciences post-1970—a field so vast, complex, and multidisciplinary it took Margaret Boden sixteen hundred pages to write its history?

As for aspirations, questions, and practices, in some ways McCulloch's clearest intellectual descendants are those who could be classified as practicing computational neuroscience—although such labels can be misleading and inaccurate.[11] Computational neuroscience, which relies heavily on computer simulation, emerged in the 1980s with its classic statement found in a paper by Terry J. Sejnowski, C. Koch, and Patricia S. Churchland: to understand the brain in terms of information processing and treat it as a computational device.[12] Like McCulloch, computational neuroscientists are interdisciplinary, are interested in the wetware of the brain, and aim to model the nervous system; yet they place much more emphasis on computation.[13] They are more scientifically conservative than McCulloch in that they ask "how" questions—how (computationally), does this part of the brain do what it does? The connections and computational processes themselves are more important than modeling higher brain functions. In contrast, McCulloch was modeling the brain, and through studying signal processing and neural circuitry, sought to make sense of the brain itself. Some later researchers, figures like Stephen Grossberg, still work in a McCulloch-esque vein: they approach the brain from a very interdisciplinary perspective, and use highly technical mathematics to tackle the brain's functioning from all angles: learning, perception, cognition, and motor control.[14]

Connectionism, the newest form of AI research, is also akin to McCulloch in terms of its focus on neural networks, and practitioners point to McCulloch's work with Pitts as an influence. But connectionists are not interested in the nervous system per se. This is more a branch of applied mathematics than brain theory. Connectionists work with idealized computational networks—sets of interconnected units. They do not seek to understand the brain itself but rather are concerned with the properties of computational processes such as pattern recognition, learning, and memory.[15]

Although their ties to McCulloch are complex, both connectionism and computational neuroscience are interdisciplinary—even transdisciplinary—endeavors. Both fields have strong ties to physics, engineering, applied mathematics, and computer science. Hypothetical modeling is a prominent method in both groups.[16] During the 1980s, connectionists and computational neuroscientists (sometimes called theoretical neuroscientists), while having a much more stable and defined disciplinary footing, faced the same sorts of disciplinary critiques that McCulloch had. Early on, computational neuroscience aroused suspicion from more empirical neuroscientists. As eminent neuroscientist Charles G. Gross told Margaret Boden:

I'm a bit dubious about the role of theory or at least of theoreticians in biology. So far, I'd suggest, no pure theoretician ever made any contribution to biology at all, and indeed they disappeared so quickly that we have all but lost their names. ... So far, theory without empirical slogging in the same head has yielded nothing.[17]

Overall, however, enough cross-disciplinary fertilization has taken place, and more and more, empirical neuroscientists discuss their questions and solutions in terms of computational processes in the brain.[18]

## 8.2   McCulloch's Identities: Context and Agency

My aim in this book has not been to get at McCulloch's essence or true character—something that I would argue is unknowable. Yet because McCulloch's identities reflect both his context and agency, his scientific life reveals important dimensions of twentieth-century American science in addition to his own idiosyncratic way of doing science.

First, McCulloch was drawn to scientific philosophy from a very early stage in his career. This interest was not simply a search for a theory of knowledge grounded in logical relations, or a theory of mind, but a view of science as a rationalizing force (particularly for making sense of diseased minds) and a preoccupation with scientific foundations. In an era where science rose to power culturally and politically, McCulloch, along with others who practiced scientific philosophy professionally, sought scientific unity. McCulloch's own brand of unity was unity of method. In the face of a very eclectic American psychiatry, McCulloch's pursuits at Yale during the 1930s were grounded in laboratory science and at the same time in discussions about the foundations of brain and behavior sciences. In these ways he began to carve out a distinctive approach to psychiatry.

When placed within this broad context, McCulloch's brand of cybernetics can be threaded into the story of his interest in scientific foundations and his work as a biological psychiatrist. McCulloch was involved in an effort to model the human sciences (and psychiatry) on the natural sciences. He looked to philosophy as a resource in this endeavor. Thus McCulloch's practices as a cybernetician—the broad, ambitious questions about the brain's overall functional organization, his development of hypothetical models with Pitts, his rhetorical strategies linking his scientific work to the problem of brain and mind and casting the brain as an information processor—can be understood, in part, as a search for a foundational, biological basis for the discipline of psychiatry. McCulloch's move to MIT may have been partly due to his loss of a niche in this world. If during the first half of the twentieth century psychiatry was eclectic, by the 1960s, psychoanalysts had all but colonized the psychiatric profession.[19] Any search for scientific foundations is really a search for the basis of a field and a project in remaking a discipline. McCulloch saw psychiatry (and neurophysiology) as "undertheorized"; he was motivated in part to cross disciplinary boundaries to bring theoretical rigor to understandings of both the brain and the diseased mind.

McCulloch's patrons also played a role in McCulloch's shifting identities and the transformation of the brain as a scientific object. Patrons can shape the contours of a scientific object: what motivates the generation of knowledge about that object, what questions are asked, what language is used to discuss it. The early philanthropic supporters of his work during the 1930s and 1940s, the Rockefeller and Macy Foundations, strove to shape the American sciences of brain and mind, and sought scientific certainty in the face of a diffuse and "unscientific" collection of therapeutic approaches to social and psychiatric problems. Much of this work was grounded in the laboratory sciences, and McCulloch's practices during this period were directed by specific, empirical questions about the functional organization of the cerebral cortex. Whether he was motivated by clinical issues surrounding epilepsy, or by a search for therapeutic techniques in treating schizophrenia or depression, by the end of the 1940s, McCulloch was a biological psychiatrist and firmly ensconced in the world of the laboratory. Here McCulloch's work was highly disciplinary and took place in the institutional world of biomedicine. Yet by the time he'd left Illinois in 1952, the institutional structure of American science had changed radically. McCulloch's primary patrons became federal units like the Office of Naval Research, who facilitated the exponential growth of multidisciplinary research groups devoted to basic science. At the RLE, McCulloch had all but abandoned clinical motivations and his role as an engineer and theoretical modeler transformed his identity as a laboratory scientist.

Similarly, the diversity of funds McCulloch received at the RLE points to the multivalent nature of the brain as a scientific object and the interdisciplinary groups interested in its function. McCulloch and his group had a complex funding model, diverse in two senses: besides the "in-house" support from the Department of Defense, by the early 1960s the group began to receive funds from other non-defense federal bodies, such as the National Institutes of Health and the National Science Foundation. With new, multidisciplinary alliances and new ideals in scientific practice being promoted by scientific patrons, the brain and its functioning were no longer the exclusive domain of medical science.

This, I would argue, is the central feature of McCulloch's scientific legacy: he helped liberate brain research from medicine—institutionally and in terms of scientific practice. By doing so, he played a crucial role in creating the modern neurosciences. The brain as a scientific object was repeatedly transformed, according to McCulloch's institutional and disciplinary milieux—from the source of epileptic convulsions to an organ at the heart of logical thought, from the basis of mental disease to an information processor. The brain emerged, post-cybernetics, as an object that transcended the divide between biology, medicine, philosophy, physics, and engineering. It became the focus of the pursuit of different kinds of knowledge: knowledge about the nature of intelligence, its functional organization, the nature of information processing and communication, and the reliability of complex systems.

What of McCulloch's own agency? McCulloch's transcending spirit was more than disciplinary or institutional. This emerges in three distinct yet overlapping ways. First, McCulloch's performances displayed a mixture of reverence and iconoclasm. He was extremely well read in classical literature and poetry, and his literary references strike one as more traditional than avant-garde. In his early days he very conventionally fashioned himself as part of a "lost generation" coming to terms with the effects of World War I. Further, his poetic form of choice was the sonnet—a restrictive, traditional form. Yet at the same time, McCulloch was unconventional and iconoclastic—ignoring disciplinary conventions, using deliberately ambiguous language, presenting himself as intellectually flamboyant. In some senses he was firmly rooted in traditions, following conventional career paths, working within traditional disciplinary boundaries; yet in many ways he ignored them.

Second, McCulloch was complex and even paradoxical in his equal ease in both practical and theoretical activities. At home in the laboratory, with over two decades of laboratory experience in studying the brain's functional organization, he also was at ease in discussions with mathematicians and engineers. Comfortable with the messiness of brain research, the use of complex instrumentation, and the challenges of clinical, psychiatric practice, McCulloch could also easily converse with scientific philosophers about logic and scientific foundations. Finally, his hands-on practical experience on the farm—surveying, designing, and building—highlights his intense interest in understanding the "go" of things: how things work in the real world. At the same time, however, he pondered universal philosophical questions that transcended time and place.

Finally, McCulloch's personal life and institutional life intersected in the generosity of spirit and openness that defined his activities both at home and in the laboratory; this was key to his legacy and to that of cybernetics. McCulloch's carefree spirit informed his very liberal, egalitarian approach to mentoring students and practicing science. Lenore Blum remembers him as "the most remarkable person one could ever hope to know … the likes of which we have never met or seen since."[20] McCulloch's students also recall him with love and admiration, not perhaps for any earthshattering discovery but for the way he made them feel. They were inspired by his unique way of doing science, of being a scientist, of inquiring about the natural world. Throughout his career, McCulloch's private and professional spheres blended together somewhat seamlessly—his homes always had an open-door policy, taking in students, collaborators, the less fortunate. While he guided both his children and students, he taught them to be self-reliant and independent. Although his penchant for rhetorical flourishes, technical details, and literary references might appear as imposing, and while some felt McCulloch deliberately went out of his way to baffle people, his son recalled: "I have seen him start a lecture looking at the audience—then pause—seeing that they were not all with him—and start again from a different take. He also gave his audience the credit of being smart—and aimed his discussion high."[21]

McCulloch died on September 24, 1969, of a heart seizure at his home in Old Lyme, at the age of 70. He had arranged for his body to be donated to Yale for research, as did Rook years later. At the time of his death, McCulloch was considered by his peers and the public as a polymath, a "confirmed eccentric" who traveled in hobnail boots, wore a brown leather jacket under his suit coat, and was known to go to sleep at 4 a.m.;

a Renaissance man who ignored disciplinary conventions yet brought vision and audacity to studies of the brain and mind.[22] In part this persona was generated by McCulloch's own self-presentation. Indeed, the image of McCulloch on *Embodiments of Mind* resembles that of a sage [figure 8.1], and the book's contents span the fields of psychiatry, engineering, neurophysiology, poetry, cybernetics, engineering, and philosophy. Capturing the image of a polymath more colorfully, in an obituary

**Embodiments of Mind** Warren S. McCulloch

**Figure 8.1**
Cover photo from *Embodiments of Mind* (1965).

for McCulloch, his friend and colleague Ralph Gerard labeled him a rebel genius.[23]

Where did this image come from? It would be tempting to say that all identities worked together coherently to bring about the rebel genius—that with such a diverse set of disciplinary experiences and practices, the rebel genius was perhaps the only identity that McCulloch *could* perform by the end of his life. Yet just as McCulloch aimed to overcome dualism in his studies of the brain, we must be anti-dualist about McCulloch—the rebel genius was not McCulloch's essence nor was it a simple construction of his late sixties milieu—although this self-reflexivity was very much of this era.[24] Rather, both context and agency come together in McCulloch's ultimate performative identity. It was a product of his way of being in the world—a coming together of his roles and varied disciplinary and institutional experiences and activities. In this sense, the rebel genius was not a composite or mosaic of the identities but a product—an intermingling of McCulloch's transdisciplinary practices, his self-fashioning, his mentoring, his personality, his construction by others. It was constructed not only by his own self-fashioning as a scientist who was single-minded, transcendent, philosophical, irreverent, and literary, but also by the recollections and observations of his contemporaries: he disregarded convention, was egalitarian, poetic, inspiring, generous, courageous, enthusiastic, curious. McCulloch was deemed rebellious because of his irreverence and critique of disciplinary convention; a genius because of his vision—not a revolutionary model of the brain, but a way of being in the world, a way of asking questions, of forming hypotheses, of transcending disciplines. As Karl Pribram observed in his review of *Embodiments of Mind*: McCulloch has "said *something* where others have mouthed banalities and nothings. ... Do not let the unconventional fool you—behind it lies much wisdom."[25]

In drawing attention to this final, complex image of McCulloch, I am not trying to impose retrospective coherence on his life. Any biography is imbued with indeterminacy. The historical vista is much larger and more complex, involving institutions, patrons, brains, animals, patients, laboratories, egos, standards, arguments, and a plethora of other factors, and thus any biography is incomplete. Ultimately, despite all his self-fashioning, in the end McCulloch would not have us search for coherence to his life. As he warned his would-be biographers:

When I am dead let no man say
That, I had lived, I had done so and so:
For I was always on an unknown way
To mine own ends, the which they could not know!

Despite his self-presentation of a life of coherence, a search for a theory of mind, McCulloch asserts his own autonomy and will, his unpredictable freedom of choice. Yet through pondering these choices, that agency, we can catch a glimpse of the rich worlds he both occupied and changed.

# Notes

## Chapter 1

1. Warren S. McCulloch, "What Is a Number, That a Man May Know It, and a Man, That He May Know a Number?" in *Embodiments of Mind* (Cambridge, MA: MIT Press, 1988), 1–18. (Originally published as The Ninth Annual Alfred Korzybski Memorial Lecture, *General Semantics Bulletin Nos. 26 and 27*, Lakeville, CT: Institute of General Semantics, 1961, 7–18.)

2. E.g., Hunter Heyck, "The Organizational Revolution and the Human Sciences," *Isis* 105, no. 1 (2014): 1–31.

3. E.g., Warren S. McCulloch, "Introductory Remarks," *Cybernetics: Circular Causal and Feedback Mechanisms in Biological and Social Systems. Transactions of the Tenth Conference April 22, 23, and 24, 1953, Princeton, N.J.*, ed. Heinz von Foerster, Margaret Mead, and Hans-Lukas Teuber (New York: Josiah Macy, Jr. Foundation, 1955), 15–18, in *Cybernetics—Kybernetik: The Macy Conferences 1946–1953*, ed. Claus Pias (Zürich-Berlin: Diaphanes, 2003), 687–688. Warren S. McCulloch, "Summary of the Points of Agreement Reached in the Previous Nine Conferences on Cybernetics," *Cybernetics*, ed. Heinz von Foerster (New York: Josiah Macy, Jr. Foundation, 1955), in Pias, *Cybernetics—Kybernetik*, 719–725. Warren S. McCulloch, "Recollections of the Many Sources of Cybernetics," in *The Collected Works of Warren S. McCulloch, Volume 1*, ed. Rook McCulloch (Salinas, CA: Intersystems, 1989), 21–49.

4. Mary Terrall, "Biography as Cultural History of Science," *Isis* 97 (2006): 306–313; Stephen Greenblatt, *Renaissance Self-Fashioning: From More to Shakespeare* (Chicago: University of Chicago Press, 1980); Mario Biagioli, *Galileo Courtier: The Practice of Science in an Age of Absolutism* (Chicago: University of Chicago Press, 1993).

5. Geof Bowker, "How to Be Universal: Some Cybernetic Strategies, 1943–1970." *Social Studies of Science* 23 (1993): 107–127.

6. Ralph W. Gerard, "Warren Sturgis McCulloch: Rebel Genius," *Transactions of the American Neurological Association* 95 (1970): 344–345.

7. Sidonie Smith and Julia Watson, *Reading Autobiography: A Guide for Interpreting Life Narratives* (Minneapolis, MN: University of Minnesota Press, 2010), 14.

8. Oren Harman, "Helical Biography and the Historical Craft: The Case of Altruism and George Price," *Journal of the History of Biology* 44 (2011): 671–691; Oren Harman, *The Price of Altruism: George Price and the Search for the Origins of Kindness* (New York: W.W. Norton, 2010).

9. Mary Terrall, "Biography as Cultural History of Science." For a look at professional identity in the context of early-twentieth-century neurology, see Delia Gavrus, "Men of Dreams and Men of Action: Neurologists, Neurosurgeons and the Performance of Professional Identity, 1925–1950," *Bulletin of the History of Medicine* 85, no. 1 (2011): 57–92.

10. An article by Thomas Hankins was a turning point (Thomas L. Hankins, "In Defence of Biography: The Use of Biography in the History of Science," *History of Science* 17 [1979]: 1–16). See Mary Terrall, *The Man Who Flattened the Earth: Maupertuis and the Sciences in the Enlightenment* (Chicago: University of Chicago Press, 2002); Mary Terrall, *Catching Nature in the Act: Réaumur and the Practice of Natural History in the Eighteenth Century* (Chicago and London: University of Chicago Press, 2014); Mary Jo Nye, *Michael Polanyi and His Generation: Origins of the Social Construction of Science* (Chicago and London: University of Chicago Press, 2011); David Aubin and Charlotte Bigg, "Neither Genius nor Context Incarnate: Norman Lockyer, Jules Janssen and the Astrophysical Self," in Thomas Söderqvist (ed.), *The History and Poetics of Scientific Biography* (Ashgate, 2007), 51–70; Sharon Ghamari-Tabrizi, *The Worlds of Herman Kahn: The Intuitive Science of Thermonuclear War* (Cambridge and London: Harvard University Press, 2005); Massimo Mazzotti, *The World of Maria Gaetana Agnesi, Mathematician of God* (Baltimore: Johns Hopkins University Press, 2012); Lorraine Daston and Otto H. Sibum, "Scientific Personae," *Science in Context* 16, no. 1 (2003): 1–18; Laura Otis, *Müller's Lab* (Oxford: Oxford University Press, 2007); Yves Gingras, "Pour une biographie sociologique," *Revue d'histoire de l'Amérique française* 45, No. 1 (2001): 123–131; Mary Jo Nye, "Scientific Biography: History of Science by Another Means?" *Isis* 97 (2006): 322–329; Mary Jo Nye, *Blackett: Physics, War, and Politics in the Twentieth Century* (Cambridge, MA: Harvard University Press, 2004); Michael Shortland and Richard Yeo (eds.), *Telling Lives in Science: Essays on Scientific Biography* (Cambridge: Cambridge University Press, 1996); Thomas Söderqvist, "Introduction: A New Look at the Genre of Scientific Biography," in Thomas Söderqvist (ed.), *The History and Poetics of Scientific Biography* (Ashgate, 2007), 1–15; Joan L. Richards, "Introduction: Fragmented Lives" *Isis* 97 (2006): 302–305.

11. Judith Butler, *Gender Trouble: Feminism and the Subversion of Identity.* (New York, London: Routledge, 1990), 112. Karen Barad, "Posthumanist Performativity: Toward

an Understanding of How Matter Comes to Matter," *Signs* 28, no. 3 (Spring 2003): 801–32; 808. I thank Natasha Meyers for alerting me to the tool of performativity.

12. Shannon Forbes, "Performative Identity Formation in Frank McCourt's *Angela's Ashes: A Memoir*," *Journal of Narrative Theory* 37, no. 3 (Fall 2007): 473–496. The notion of identity I am using here aligns with the notion of a scientific persona, which lies at the intersection of the individual and the social institution. Personae, on this reading, are not masks of a true stable self, but rather, according to Daston and Sibum, help one attain selfhood and "create possibilities of being in the world." Lorraine Daston and Otto Sibum, "Scientific Personae," *Science in Context* 16, no. 1 (2003): 1–18, 3–4.

13. Barad, "Posthumanist Performativity," 807. Like Barad's notion of agential realism: just as gender is not an inherent feature of a subject, the body is not a "passive blank slate" on which history or culture makes its mark. Karen Barad, *Meeting the Universe Halfway: Quantum Physics and the Entanglement of Matter and Meaning* (Duke University Press, 2007), 60.

14. In many ways my analysis of McCulloch aligns with Jan Golinski's treatment of another self-fashioning figure—Humphry Davy (Jan Golinski, "Humphry Davy: The Experimental Self," *Eighteenth-Century Studies* 45, no. 1 (Fall 2011) 15–28. See also Jan Golinski, *The Experimental Self: Humphry Davy and the Making of a Man of Science* (Chicago: University of Chicago Press, 2016). Golinski examines how personal identities vis-à-vis emerging disciplines provided Davy with a means of articulating his own subjectivity. Yet there are important contrasts with McCulloch's case: McCulloch did not reflect on his own identity in the same way as Davy, disciplinary structures were just coming into being in the early nineteenth century (whereas disciplines were well established in McCulloch's time), and Davy's life was much more public than McCulloch's. In contrast, McCulloch *grew into* a public figure.

15. See James Clifford, "'Hanging up Looking Glasses at Odd Corners': Ethnobiographical Prospects," in *Studies in Biography*, ed. D. Aaron (Cambridge, MA: Harvard University Press, 1978), 41–56.

16. In addressing the question of how McCulloch's work in the laboratory and the clinic influenced his philosophical approach to the mind, I am informed by the work of Andrew Pickering, who shows how much of the work of British cyberneticians was informed by their disciplinary grounding in psychiatry. Andrew Pickering. *The Cybernetic Brain: Sketches of Another Future* (Chicago: University of Chicago Press, 2010).

17. On the disciplinary structure of the sciences, see Charles Rosenberg, "Toward an Ecology of Knowledge: On Discipline, Context, and History," in Alexandra Oleson and John Voss (eds.) *The Organization of Knowledge in Modern America, 1860–1920* (Baltimore, MD: Johns Hopkins University Press, 1979), 440–455; Robert E. Kohler,

*From Medical Chemistry to Biochemistry: The Making of a Biomedical Discipline* (Cambridge University Press, 1982); Julie Thompson Klein, *Crossing Boundaries: Knowledge, Disciplinarities, and Interdisciplinarities* (Charlottesville & London: University Press of Virginia, 1996); James M. M. Good, "Disciplining Social Psychology: A Case Study of Boundary Relations in the History of the Human Sciences," *Journal of the History of the Behavioral Sciences* 36, no. 4 (Fall 2000): 383–403; Timothy Lenoir (ed.), *Instituting Science: The Cultural Production of Scientific Disciplines* (Stanford: Stanford University Press, 1997), Chapter 2.

18. Rosenberg, "Toward an Ecology of Knowledge," 443. See also Steve Fuller, "Disciplinary Boundaries and the Rhetoric of the Social Sciences," *Poetics Today* 12, no. 2 (Summer 1991): 301–325, 302.

19. Good, "Disciplining Social Psychology."

20. Helga Nowotny, "The Potential of Transdisciplinarity." First published in *Inter-disciplines*, http://www.inter-disciplines.org, May 2006, accessed 15 August 2014; Jill A. Lazenby, "Climates of Collaboration: Interdisciplinary Science and Social Identity", Ph.D. Dissertation, University of Toronto, 2003; Julie Thompson Klein, *Crossing Boundaries*; Fuller, "Disciplinary Boundaries"; Lenoir, *Instituting Science*. On interdisciplinarity and transdisciplinarity in late-twentieth-century human sciences, see Hunter Crowther-Heyck, *Herbert A. Simon: The Bounds of Reason in Modern America* (Johns Hopkins University Press, 2005); Hunter Crowther-Heyck, "Patrons of the Revolution: Ideals and Institutions in Postwar Behavioral Science," *Isis*, vol. 97, no. 3 (September 2006): 420–446; Jamie Cohen-Cole, *The Open Mind: Cold War Politics and the Sciences of Human Nature* (Chicago: University of Chicago Press, 2014); Jamie Cohen-Cole, "Instituting the Science of Mind: Intellectual Economies and Disciplinary Exchange at Harvard's Center for Cognitive Studies," *British Journal for the History of Science* 40, no. 4 (2007): 567–597.

21. On the renegotiation of boundaries and the strategies behind such practices, see Thomas Gieryn, *Cultural Boundaries of Science: Credibility on the Line* (Chicago: University of Chicago Press, 1999).

22. See Hans-Jörg Rheinberger, *Toward a History of Epistemic Things: Synthesizing Proteins in the Test Tube* (Stanford, CA: Stanford University Press, 1997), 28, on "materializing questions."

23. Tara H. Abraham, "From Theory to Data: Representing Neurons in the 1940s," *Biology and Philosophy* 18, no. 3 (2003): 415–426. Tara H. Abraham, "(Physio)logical Circuits: The Intellectual Origins of the McCulloch–Pitts Neural Networks," *Journal of the History of the Behavioral Sciences* 38, no. 1 (2002): 3–25.

24. Hans-Jörg Rheinberger, "Experiment, Difference, and Writing. Part 1. Tracing Protein Synthesis. Part 2. The Laboratory Production of Transfer RNA," *Studies in History and Philosophy of Science* 23 (1992): 305–31, 389–422; Robert E. Kohler, "Sys-

tems of Production: Drosophila, Neurospora, and Biochemical Genetics," *Historical Studies in the Physical and Biological Sciences* 22 (1991): 87–130.

25. Angela N. H. Creager, *The Life of a Virus: Tobacco Mosaic Virus as an Experimental Model, 1930–1965* (Chicago: University of Chicago Press, 2002), Chapter 8; Robert E. Kohler, *Lords of the Fly:* Drosophila *Genetics and the Experimental Life* (Chicago: University of Chicago Press, 1994); Bruno Latour and Steve Woolgar, *Laboratory Life: The Construction of Scientific Facts*, 2nd ed. (Princeton, NJ: Princeton University Press, [1979], 1986). See for the medical context, Ilana Löwy, "Experimental Systems and Clinical Practices: Tumor Immunology and Cancer Immunotherapy, 1895–1980," *Journal of the History of Biology* 27 (1994): 403–35; Joan Fujimura, "Standardizing Practices: A Socio-History of Experimental Systems in Classical Genetic and Virological Cancer Research, ca.1920–1978," *History and Philosophy of the Life Sciences* 18 (1996): 3–54.

26. Stathis Psillios, "An Explorer upon Untrodden Ground: Peirce on Abduction," in *The Handbook of the History of Logic, Volume 10: Inductive Logic*, ed. Dov M. Gabbay, Stephan Hartmann, and John Woods (North Holland, 2011), 117–151.

27. McCulloch, "What Is a Number," 5. For neuroscientist Peter Cariani, this amounts to a form of "reverse-engineering": postulating a mechanism by which the brain may carry out a function and comparing it to the observed behavior of biological brains. Peter Cariani, Personal Communication, May 2015.

28. Sabina Leonelli, "What Is in a Model? Combining Theoretical and Material Models to Develop Intelligible Theories. In *Modeling Biology: Structures, Behaviors, Evolution*, ed. Manfred D. Laubichler and G. M. Müller, 15–36. (Cambridge, MA: MIT Press, 2007); Mary S. Morgan and Margaret Morrison (eds.), *Models as Mediators: Perspectives on Natural and Social Science* (Cambridge, MA: Cambridge University Press, 1999); Angela N. H. Creager, M. Norton Wise, and Elizabeth Lunbeck (eds.), *Science Without Laws: Model Systems, Cases, Exemplary Narratives* (Durham, NC: Duke University Press, 2007).

29. E.g., Ilana Löwy, "The Strength of Loose Concepts—Boundary Objects, Federative Experimental Strategies, and Disciplinary Growth: The Case of Immunology," *History of Science* 30 (1992): 371–396.

30. Peter Galison, *Image and Logic: A Material Culture of Microphysics* (Chicago: University of Chicago Press, 1997).

31. Leah Ceccarelli, *Shaping Science with Rhetoric: The Cases of Dobzhansky, Schrödinger, and Wilson* (Chicago and London: University of Chicago Press, 2001).

32. Ceccarelli, *Shaping Science with Rhetoric*, 5; Leah Ceccarelli, "Polysemy: Multiple Meanings in Rhetorical Criticism," *Quarterly Journal of Speech* 84, no. 4 (1998): 395–415.

33. Ceccarelli, *Shaping Science with Rhetoric*, 169–170.

34. A. Hunter Dupree, "The Great Instauration of 1940: The Organization of Scientific Research for War," in *The Twentieth-Century Sciences: Studies in the Biography of Ideas*, ed. Gerald Holton (New York: Norton, 1972), 443–467.

35. E.g., Toby A. Appel, *Shaping Biology: The National Science Foundation and American Biological Research, 1945–1975* (Baltimore and London: Johns Hopkins University Press, 2000).

36. Hunter Crowther-Heyck, "Patrons of the Revolution"; Mark Solovey, *Shaky Foundations: The Politics-Patronage-Social Science Nexus in Cold War America* (New Brunswick, NJ: Rutgers University Press, 2013); Jamie Cohen-Cole, *The Open Mind*.

37. Paul N. Edwards, *The Closed World: Computers and the Politics of Discourse in Cold War America* (Cambridge, MA: MIT Press, 1996); N. Katherine Hayles, *How We Became Posthuman: Virtual Bodies in Cybernetics, Literature, and Informatics* (Chicago: University of Chicago Press, 1999).

38. Steve J. Heims, *Constructing a Social Science for Postwar America: The Cybernetics Group, 1946–1953*. Cambridge, MA: MIT Press, 1993); Paul N. Edwards, *The Closed World*.

39. E.g., Gualtiero Piccinini, "The First Computational Theory of Mind and Brain: A Close Look at McCulloch and Pitts's 'Logical Calculus of Ideas Immanent in Nervous Activity,'" *Synthese* 141, no. 2 (2004): 175–215; Kenneth Aizawa and Mark Schlatter, "Walter Pitts and 'A Logical Calculus,'" *Synthese*, 162 (2007), 235–250.

40. Jean-Pierre Dupuy, *The Mechanization of the Mind: On the Origins of Cognitive Science*, trans. M. B. DeBevoise (Princeton, NJ: Princeton University Press, 2000). (Original work published 1994.)

41. Lily E. Kay, "From Logical Neurons to Poetic Embodiments of Mind: Warren S. McCulloch's Project in Neuroscience," *Science in Context* 14, no. 4 (2001): 591–614.

42. See, for example, Bernard J. Baars, *The Cognitive Revolution in Psychology* (New York and London: Guilford Press, 1986); Margaret Boden, *Mind as Machine: A History of Cognitive Science*, 2 vols. (Oxford: Oxford University Press, 2006); Howard Gardner, *The Mind's New Science* (New York: Basic Books, 1985); J. Leiber, *An Invitation to Cognitive Science* (Oxford: Basil Blackwell, 1991); Christina E. Erneling and David Martel Johnson, *The Mind as a Scientific Object: Between Brain and Culture* (Oxford University Press, 1997). However, a notable exception is T. Leahy, "The Mythical Revolutions of American Psychology," *American Psychologist* 47 (1992): 308–18. For a nuanced picture of the changes taking place in psychology during this period, see Jamie Cohen-Cole, "The Reflexivity of Cognitive Science: The Scientist as a Model of Human Nature," *History of the Human Sciences* 18, no. 4 (2005): 107–139.

43. Roger Smith, *The Human Sciences* (New York: W.W. Norton, 1997), 832.

44. Roger Smith, "The History of Psychological Categories," *Studies in History and Philosophy of Science Part C: Studies in History and Philosophy of Biological and Biomedical Sciences* 36 (2005): 55–94; Kurt Danziger, *Naming the Mind: How Psychology Found Its Language* (London: Sage, 1997) presents a similar position on scientific categories.

45. See, e.g., Kurt Danziger, *Constructing the Subject: Historical Origins of Psychological Research* (Cambridge: Cambridge University Press, 1990); Kurt Danziger, "Social Context and Investigative Practice in Early Twentieth-Century Psychology," in *Psychology in Twentieth-Century Thought and Society*, ed. Mitchell G. Ash and William R. Woodward (Cambridge: Cambridge University Press, 1987), 13–33. I adopt Danziger's definition of investigative practice as "a social practice, in the sense that the individual investigator acts within a framework determined by the potential consumers of his or her research and by the traditions of acceptable practice prevailing in the field" (1990, 4).

46. Peter Galison, "The Ontology of the Enemy: Norbert Wiener and the Cybernetic Vision," *Critical Inquiry* 21, no. 1 (1994): 228–266; David A. Mindell, *Between Human and Machine: Feedback, Control, and Computing Before Cybernetics* (Baltimore and London: Johns Hopkins University Press, 2002); Andrew Pickering, "Cyborg History and the World War II Regime," *Perspectives on Science* 3 (1995): 1–48; Eden Medina, *Cybernetic Revolutionaries: Technology and Politics in Allende's Chile* (Cambridge, MA: MIT Press, 2011); Ronald R. Kline, *The Cybernetics Moment: Or Why We Call Our Age the Information Age* (Baltimore, MD: Johns Hopkins University Press, 2015).

47. Steve J. Heims, *Constructing a Social Science for Postwar America*; Philip Mirowski, *Machine Dreams: Economics Becomes a Cyborg Science* (Cambridge University Press, 2002).

48. Paul N. Edwards, *The Closed World*.

49. N. Katherine Hayles, *How We Became Posthuman*.

50. Lily E. Kay, *Who Wrote the Book of Life? A History of the Genetic Code* (Stanford: Stanford University Press, 2000); Evelyn Fox Keller, *Making Sense of Life: Explaining Biological Development with Models, Metaphors, and Machines* (Cambridge, MA: Harvard University Press, 2002); Evelyn Fox Keller, *Refiguring Life: Metaphors of Twentieth-Century Biology* (New York: Columbia University Press, 1995).

51. Peter Asaro, "Working Models and the Synthetic Method: Electronic Brains as Mediators between Neurons and Behavior," *Science Studies* 19, no. 1 (2006): 12–34.

52. Andrew Pickering, *The Cybernetic Brain*.

53. Eden Medina, *Cybernetic Revolutionaries*.

54. Michael Hagner and Cornelius Borck, "Mindful Practices: On the Neurosciences in the Twentieth Century," *Science in Context* 14, no. 4 (2001): 507–510.

55. L. S. Jacyna, *Lost Words: Narratives of Language and the Brain, 1825–1926* (Princeton: Princeton University Press, 2000); Andrew Shail and Laura Salisbury, eds., *Neurology and Modernity: A Cultural History of Nervous Systems, 1800–1950* (Palgrave Macmillan, 2010); Kenton Kroker, *The Sleep of Others and the Transformations of Sleep Research* (Toronto: University of Toronto Press, 2007). See also on brain surgery, Alison Winter, *Memory: Fragments of a Modern History* (Chicago: University of Chicago Press, 2012), chapter 4.

56. E.g., Horace W. Magoun and Louise H. Marshall (eds.), *American Neuroscience in the Twentieth Century: Confluence of the Neural, Behavioral, and Communicative Streams* (Swets and Zeitlinger, 2002); Frederic Worden, Judith P. Swazey, and George Adelman (eds.), *The Neurosciences: Paths of Discovery* (Cambridge, MA: MIT Press, 1975); Gordon M. Shepherd, *Creating Modern Neuroscience: The Revolutionary 1950s* (New York: Oxford University Press, 2010); Charles G. Gross, *A Hole in the Head: More Tales in the History of Neuroscience* (Cambridge, MA: MIT Press, 2009); Stanley Finger, *Origins of Neuroscience: A History of Explorations into Brain Function* (New York, Oxford: Oxford University Press, 1994).

57. E.g., Robert M. Young, *Mind, Brain, and Adaptation: Cerebral Localization and Its Biological Context from Gall to Ferrier* (Oxford: Clarendon Press, 1970); Susan Leigh Star, *Regions of the Mind: Brain Research and the Quest for Scientific Certainty* (Stanford: Stanford University Press, 1989); Anne Harrington, *Medicine, Mind, and the Double Brain: A Study in Nineteenth-Century Thought* (Princeton: Princeton University Press, 1987).

58. Ronald Kline, "Where Are the Cyborgs in Cybernetics?" *Social Studies of Science*, 33 (June 2009): 331–362.

## Chapter 2

1. William Eckhardt, "Changing Concerns in Peace Research and Education," *Bulletin of Peace Proposals* 3 (1974): 280. The term "transdisciplinary" was apparently coined at an a meeting of the Organization for Economic Co-operation and Development in Nice in 1970 and developed further by Erich Jantsch in 1972, when he described an ideal of a "cybernetic university." See Andrew Barry and Georgina Born, "Interdisciplinarity: Reconfigurations of the Social and Natural Sciences," in *Interdisciplinarity: Reconfigurations of the Social and Natural Sciences*, ed. Andrew Barry and Georgina Born (London and New York: Routledge, 2013), 1–56, 8. See also Helga Nowotny, "The Potential of Transdisciplinarity," first published in *Interdisciplines*, http://www.inter-disciplines.org, May 2006 (accessed 15 August 2014); and Julie Thompson Klein, ed., *Transdisciplinarity: Joint Problem Solving Among Science, Technology, and Society* (Basel: Birkhauser, 2001) for definitions of transdisciplinarity.

2. Julie Thompson Klein, "The Transdisciplinary Moment(um)," *Integral Review* 9, no. 2 (June 2013): 189–199.

3. Richard Carter, "William Carlos Williams (1883–1963): Physician-Writer and 'Godfather of Avant-Garde Poetry,'" *Annals of Thoracic Surgery* 67, (1999): 1512–1517.

4. See Lisa Keränen, *Scientific Characters: Rhetoric, Politics, and Trust in Breast Cancer Research* (Tuscaloosa, AL: University of Alabama Press, 2010), on ethos, persona, and voice as modeling "character" in medical research.

5. "James W. McCulloch (Notes by daughter, Margaret C. McCulloch, for David Sears McCulloch, Christmas, 1966.)" Box 5, Warren S. McCulloch Family Photographs, Series VIII, WSM Papers, APS.

6. Willard D. Litt (ed.) *History of the Class of Nineteen Hundred and Twenty-One* (New Haven, CT: Tuttle, Morehouse, and Taylor, 1921), 222.

7. Wedding Invitation, Group 299, Box 32, Folder 523—Correspondence Mc 1922–1924 Oct 31, Series I, Anson Phelps-Stokes Family Papers (APS Papers), YUL.

8. Anson Phelps-Stokes to Miss Macdonald, 29 November 1927, Group 299, Box 32, Folder 527—Correspondence Mc 1927 Jul 16–1928 Dec 31, Series I, APS Papers, YUL.

9. Margaret McCulloch, "Warren—Chapter II—School Days," 1973, Box 4, Warren S. McCulloch Family Photographs, Series VIII, WSM Papers, APS. Beards was a girls' school but accepted both boys and girls for grades up to 3.

10. Margaret McCulloch, "Warren—Chapter II—School Days."

11. Margaret McCulloch, "Warren McCulloch—First Character" circa 1973, Box 4, Warren S. McCulloch Family Photographs, Series VIII, WSM Papers, APS.

12. David McCulloch, Personal Communication, 4 February 2015.

13. Rook McCulloch, "Foreword," in *The Collected Works of Warren S. McCulloch, Volume I*, ed. Rook McCulloch (Salinas, CA: Intersystems, 1989), 1–6, p. 1.

14. See Rufus M. Jones, *Haverford College: A History and Interpretation* (Macmillan, 1933); Gregory Kannerstein (ed.), *The Spirit and the Intellect: Haverford College 1833–1983* (Haverford, PA: Haverford College, 1983).

15. Diana Alten, "Rufus Jones and the American Friend: A Quest for Unity," *Quaker History*, 74, No. 1 (Spring 1985): 41–48.

16. Alten, "Rufus Jones."

17. Warren S. McCulloch, *The Natural Fit* (Chicago: Chicago Literary Club, 1959), 5.

18. National Board of Medical Examiners (n.d.), Copy of Certificate of Premedical College Work for Warren McCulloch, in Folder "McCulloch, Warren S. (Warren Sturgis), 1941–1949, Curriculum vitae," Series II, WSM Papers, APS, McCulloch, "Recollections of the many sources of cybernetics."

19. Rook McCulloch, "Foreword," 2.

20. Margaret C. McCulloch "Postscript to School Days," circa 1973, Box 4, Warren S. McCulloch Family Photographs, Series VIII, WSM Papers, APS.

21. US World War I Draft Registration Cards, 1917–1918, The National Archives, Provo, UT: Ancestry.com.

22. E.g., Lynn Dumenil, *The Modern Temper: American Culture and Society in the 1920s* (New York: Hill and Wang, 1995).

23. Warren Sturgis McCulloch, Curriculum Vitae, Folder "McCulloch, Warren S. (Warren Sturgis), 1941–1949, Curriculum vitae," Series II, WSM Papers, APS; Litt, *History of the Class of Nineteen Hundred and Twenty-One*, 222.

24. John R. Thelin, *A History of American Higher Education*, 2nd Edition (Baltimore, MD: Johns Hopkins University Press, 2011), 200.

25. Brooks Mather Kelley, *Yale: A History* (New Haven and London: Yale University Press, 1974), 349.

26. Michael D. Besch, *A Navy Second to None: The History of U.S. Naval Training in World War I* (Westport, CT and London: Greenwood Press, 2002), 180.

27. Kelley, *Yale: A History*, 353.

28. Litt, *History of the Class of Nineteen Hundred and Twenty-One*, 359–363.

29. *Yale Alumni Weekly*, 11 October 1918, quoted in Kelley, *Yale: A History*, 353.

30. Paula Fass, *The Damned and the Beautiful: American Youth in the 1920s* (Oxford: Oxford University Press, 1979), 5.

31. Fass, *The Damned and the Beautiful*, 6, 19.

32. Fass, *The Damned and the Beautiful*, 330.

33. Litt, *History of the Class of Nineteen Hundred and Twenty-One*, 69.

34. David McCulloch, Personal Communication, 4 February 2015.

35. Litt, *History of the Class of Nineteen Hundred and Twenty-One*, 367.

36. Litt, *History of the Class of Nineteen Hundred and Twenty-One*, 1.

37. A collection of his poetry was read before the Chicago Literary Club in 1957, published in 1959 as *The Natural Fit*.

38. David Graham and Kate Sontag, "Containing Multitudes," in *After Confession: Poetry as Autobiography*, ed. Kate Sontag and David Graham (Saint Paul, MN: Graywolf Press, 2001), 3–8.

39. McCulloch, *The Natural Fit*, 5.

40. Paul Oppenheimer, "The Origin of the Sonnet," *Comparative Literature* 34, no. 4 (Fall 1982): 289–304.

41. McCulloch, *The Natural Fit*, 5.

42. Fass, *The Damned and the Beautiful*, 18–19.

43. Fass, *The Damned and the Beautiful*, 28.

44. Fass, *The Damned and the Beautiful*, 28.

45. McCulloch, *The Natural Fit*, 33.

46. McCulloch, *The Natural Fit*, 29.

47. Fass, *The Damned and the Beautiful*, 30.

48. Fass, *The Damned and the Beautiful*, 383n31.

49. Ben B. Lindsey and Wainwright Evans, *The Revolt of Modern Youth* (New York: Boni & Liveright, 1925), 54.

50. Lindsey and Evans, *The Revolt of Modern Youth*, 157.

51. McCulloch, *The Natural Fit*, 32.

52. John Tinterman Newcomb, "The Footprint of the Twentieth Century: American Skyscrapers and Modernist Poems," *Modernism/Modernity* 10, no. 1 (January 2003): 97–125.

53. McCulloch, *The Natural Fit*, 31.

54. Dumenil, *The Modern Temper*, 151–52.

55. Frederick Lewis Allen, *Only Yesterday: An Informal History of the Nineteen-Twenties* (New York: Harper, 1951), 197; Dumenil, *The Modern Temper*, Chapter IV.

56. Andrew Jewett, *Science, Democracy, and the American University: From the Civil War to the Cold War* (Cambridge: Cambridge University Press, 2012), 10.

57. Bruce Kuklick, *A History of Philosophy in America, 1720–2000* (Oxford: Clarendon Press, 2001), 100.

58. "Certificate of Premedical College Work," WSM Papers, APS.

59. Bruce Kuklick, "Philosophy at Yale in the century after Darwin," *History of Philosophy Quarterly* 21, no.3 (2004): 313–336; 318–319.

60. Bruce Kuklick, *The Rise of American Philosophy: Cambridge, Massachusetts 1860–1930* (New Haven and London: Yale University Press, 1977), 566.

61. This was not quite philosophy of science, whose emergence as a discipline would wait until the 1930s influx of émigré logical empiricists, e.g., Rudolf Carnap at Chicago and UCLA; Herbert Feigl at Minnesota.

62. Although it is unclear from which part of McCulloch's life his copy of Peirce's *Collected Works* dates from, he had placed bookmarks in the section on logic and quantity.

63. Laurence D. Smith, *Behaviorism and Logical Positivism: A Reassessment of the Alliance* (Stanford, CA: Stanford University Press, 1986), 38.

64. Bertrand Russell, "Logical Atomism" in *Logical Positivism*, ed. Alfred J. Ayer (Glencoe, IL: Free Press, 1959), 31–50; Bertrand Russell, *Analysis of Mind* (London: George Allen and Unwin; New York: Macmillan, 1921).

65. Alan Richardson, "Philosophy of Science in America," in Cheryl Misak (ed.), *The Oxford Handbook of American Philosophy* (Oxford Handbooks Online, Oxford University Press, 2013), 1–53, pp. 6–7. DOI: 10.1093/oxfordhb/9780199219315.003.0017, retrieved 29 July 2013; Francesca Bordogna, *William James at the Boundaries: Philosophy, Science, and the Geography of Knowledge* (Chicago and London: University of Chicago Press, 2008), Chapter 1.

66. E.g., Alan Richardson, "The Geometry of Knowledge: Lewis, Becker, Carnap, and the Formalization of Philosophy in the 1920s," *Studies in the History and Philosophy of Science* 34, (2003): 165–182.

67. Richardson, "Philosophy of Science in America," 11.

68. Kuklick, *A History of Philosophy in America*, 112.

69. Kuklick, *A History of Philosophy in America*, 151; Daniel J. Wilson, *Science, Community, and the Transformation of American Philosophy, 1860–1930* (Chicago and London: University of Chicago Press, 1990), Chapter 1.

70. Wilson, *Science, Community, and the Transformation of American Philosophy*, 32.

71. Bordogna, *William James at the Boundaries*.

72. Bordogna, *William James at the Boundaries*, 79–84.

73. Smith, *Behaviorism and Logical Positivism*, 3.

74. Bruce Kuklick, "Philosophy at Yale," 317.

75. Margaret C. McCulloch "Postscript to School Days."

76. Hambidge's lectures were published in 1926 as *Elements of Dynamic Symmetry* (New York: Brentano).

77. Harold J. McWhinnie, "A Biological Basis for the Golden Section in Art and Design," *Leonardo* 19, no. 3 (1985): 241–245.

78. Mark Andrew White, "Hambidge, Jay," http://www.anb.org.myaccess.library.utoronto.ca/articles/17/1700364.html, *American National Biography Online* Feb. 2000. Accessed 29 July 2013.

79. For a review of this work, see Christopher D. Green, "All That Glitters: A Review of Psychological Research on the Aesthetics of the Golden Section," *Perception* 24 (1995): 937–968.

80. John G. Benjafield, "The Golden Section and American Psychology, 1892–1938," *Journal of the History of the Behaviorial Sciences* 46, no. 1 (2010): 52–71.

81. Elspeth H. Brown, *The Corporate Eye: Photography and the Rationalization of American Commercial Culture* (Baltimore: Johns Hopkins University Press, 2005), 168.

82. Warren S. McCulloch, "Dynamic Symmetry" 8 June 1921; Folder "Dynamic Symmetry," Series II, WSM Papers, APS.

83. McCulloch, "Dynamic Symmetry."

84. Litt, *History of the Class of Nineteen Hundred and Twenty-One*, 222.

85. Warren S. McCulloch, "A Preference for Related Areas, Submitted in Partial Fulfillment of the Requirement for the Degree of Master of Arts in the Faculty of Philosophy, Columbia University," n.d., Folder "Dynamic Symmetry," Series II, WSM Papers, APS.

86. "Albert Poffenberger, 92; Professor at Columbia," *New York Times*, 27 December 1977, p. 38.

87. Roland Marchand, *Advertising the American Dream: Making Way for Modernity, 1920–1940* (Berkeley: University of California Press, 1985), 235–236.

88. Frederick C. Thorne, "Reflections on the Golden Age of Columbia Psychology," *Journal of the History of the Behavioral Sciences* 12 (1976): 159–165.

89. Michael M. Sokal, "James McKeen Cattell and American Psychology in the 1920s," in *Explorations in the History of Psychology in the United States*, ed. Josef Brožek (Lewisburg: Bucknell University Press; London: Associated University Presses, 1984), 273–323; 278–280.

90. Thorne, "Reflections," 159.

91. Thorne, "Reflections," 162.

92. William G. Rothstein, *American Medical Schools and the Practice of Medicine: A History* (New York: Oxford University Press, 1987), 142.

93. "Certificate of Premedical College Work," WSM Papers, APS.

94. Interview with Taffy Holland and David McCulloch, Old Lyme, Connecticut, 13 July 2002; David McCulloch, Personal Communication, 4 February 2015.

95. Alex Andrew, "Tribute to the Life and Work of Rook McCulloch," *Kybernetes* 22, no. 3 (1993): 4. Rook's mother, Mina Boehm Metzger (1877–1975), studied art under Arshile Gorky at the Grand Central Art School during the 1930s and

collected his works, many of which are now at the Whistler House Museum of Art in Lowell, MA.

96. Horace Coon, *Columbia: Colossus on the Hudson* (New York: E.P. Dutton & Co., 1947), 263.

97. "III.—The medical school of Columbia University: College of Physicians and Surgeons," *The Lancet* 213 (5 January 1929), e9-e11.

98. "National Board of Medical Examiners: Certificate of Medical Education," 10 July 1925, WSM Papers, APS.

99. "Certificate of Medical Education," 13 November 1928, WSM Papers.

100. McCulloch "Recollections," 30.

101. Harry M. Marks, *The Progress of Experiment: Science and Therapeutic Reform in the United States, 1900–1990* (Cambridge University Press, 1997). See also Andrew Cunningham and Perry Williams (eds.), *The Laboratory Revolution in Medicine* (Cambridge University Press, 1992).

102. Christopher G. Goertz, "Early American Professorships in Neurology," *Annals of Neurology* (1996) 40(2): 258–263; Bonnie Ellen Blustein, "New York Neurologists and the Specialization of American Medicine," *Bulletin of the History of Medicine* (Summer 1979) 53, 2: 170–183; Stephen T. Casper, "A Revisionist History of American Neurology," *Brain* 133 (2010): 638–642; Stephen T. Casper, "Conjunctures in Anglo-American Neurology: Lewis H. Weed and Johns Hopkins Neurology, 1917–1942," *Bulletin of the History of Medicine* 82 (2008): 646–671. For a thorough treatment of debates about identity within neurology, particularly neurosurgery, during this period, see Delia Gavrus, "Men of Dreams and Men of Action: Neurologists, Neurosurgeons, and the Performance of Professional identity," *Bulletin of the History of Medicine* (2011) 85: 57–92.

103. Gavrus, "Men of Dreams," 73.

104. John Hughlings Jackson, "On the Anatomical and Physiological Localization of Movements in the Brain," *The (London) Lancet*, no. 4 (April 1873): 197–201 [Part I]; no. 5 (May): 245–48 [Part 2].

105. Ellen Dwyer, "Neurological Patients as Experimental Subjects: Epilepsy Studies in the United States," in *The Neurological Patient in History*, ed. L. S. Jacyna and Stephen T. Casper (Rochester, NY: Rochester University Press, 2012), 44–60, 45.

106. Alejandra C. Laszlo, "Physiology of the Future: Institutional Styles at Columbia and Harvard," in *Physiology in the American Context, 1850–1940*, ed. Gerald L. Geison (Baltimore, MD: American Physiological Society, 1987), 67–96.

107. F. H. Pike, Charles A. Elsberg, W. S. McCulloch, and A. Rizzolo, "Some Observations on Experimentally Produced Convulsions," *American Journal of Psychiatry* 9

(1929): 259–283; F. H. Pike, C. A. Eslberg, W. S. McCulloch, and M. N. Chappell, "The Problem of Localization in Experimentally Induced Convulsions," *Archives of Neurology and Psychiatry* 23, no. 5 (1930): 847–868; 859–860. Tonic-clonic seizures are referred to as "grand mal" seizures.

108. Pike et al., "Some Observations on Experimentally Produced Convulsions."

109. Pike et al., "The Problem of Localization in Experimentally Induced Convulsions."

110. Sandra Opdycke, *No One Was Turned Away: The Role of Public Hospitals in New York City Since 1900* (New York, Oxford: Oxford University Press, 1999), 45.

111. See, for example, S. Bernard Wortis and Warren S. McCulloch, "Head Injuries: An Experimental Study," *Archives of Surgery* 25 (1932): 529–543. Bellevue was affiliated with the New York University medical school, and eventually Wortis became an eminent figure at Bellevue, deemed largely responsible for the "golden age" of NYU/Bellevue psychiatry. See Morris Herman, "S. Bernard Wortis 1904–1969," *Transactions of the American Neurological Association* 95 (1970): 343–344.

112. This drug is not to be confused with other bromides, which were commonly used as anti-epileptic treatments during the early twentieth century. See Walter J. Friedlander, "The Rise and Fall of Bromide Therapy in Epilepsy," *Archives of Neurology* 57, no. 12 (2000): 1782–1785.

113. Ellen Dwyer, "Toward New Narratives of Twentieth-Century Medicine," *Bulletin of the History of Medicine* 74, no. 4 (Winter 2000): 786–93, 789. Dwyer cites the 1922 and 1929 meetings of the Association for Research in Nervous and Mental Disease, the proceedings of which were published in *Epilepsy and the Convulsive State: An Investigation of the Most Recent Advances, Parts 1 and 2* (Baltimore: Williams and Wilkins, 1931).

114. Dwyer, "Toward New Narratives"; Gavrus, "Men of Dreams"; Dwyer, "Neurological Patients as Experimental Subjects," 45; Jack Pressman, *Last Resort: Psychosurgery and the Limits of Medicine* (Cambridge: Cambridge University Press, 1998).

115. Andrew, "Tribute."

116. Interview with Taffy Holland and David McCulloch, 13 July 2002; Andrew, "Tribute."

117. McCulloch Farm—Farm History, http://www.whippoorwillmorgans.com/html/farmhistory-orig.html. Accessed 10 July 2015.

118. David McCulloch, Personal Communication, 4 February 2015.

119. Russell E. Blaisdell, "The Rockland State Hospital," *Psychiatric Quarterly* 5, no.1 (1931): 193–195. Construction began on Rockland in 1927, and it opened in 1931.

120. Frederick W. Parsons, "The Problem of Mental Hygiene in New York State," *Psychiatric Quarterly* (1927) 1(3): 271–275.

121. Nancy Tomes, "Introduction," *Bulletin of the History of Medicine* 74, no. 4 (Winter 2000): 773–777.

122. George Weisz, *Divide and Conquer: A Comparative History of Medical Specialization* (Oxford University Press, 2006).

123. Pressman, *Last Resort*, Chapter 1. On the evolving relationship between neurology and psychiatry, see L. Casamajor, "Notes for an Intimate History of Neurology and Psychiatry," in *American Journal of Nervous and Mental Disease* 98 (1943): 600–608; Bonnie Ellen Blustein, "New York Neurologists"; Bonnie Ellen Blustein, "'A Hollow Square of Psychological Science': American Neurologists and Psychiatrists in Conflict," in *Madhouses, Mad-Doctors, and Madmen*, ed. Andrew Scull (Philadelphia: University of Pennsylvania Press, 1981), 241–270.

124. Bonnie Ellen Blustein. "'A Hollow Square of Psychological Science.'"

125. Gerald N. Grob, *Mental Illness and American Society: 1870–1940* (Princeton, NJ: Princeton University Press, 1983), xi.

126. Grob, *Mental Illness*, Chapter 3.

127. Grob, *Mental Illness*, Chapter 5.

128. Hans Pols, "'Beyond the Clinical Frontiers': The American Mental Hygiene Movement, 1910–1945," in Volker Roelke, Paul J.Weindling, and Louise Westwood (eds.), *International Relations in Psychiatry: Britain, Germany, and the United States to World War II* (Rochester, NY: University of Rochester Press, 2013), 111–133.

129. Nathan G. Hale, Jr., *The Rise and Crisis of Psychoanalysis in the United States: Freud and the Americans 1917–1995* (New York, Oxford: Oxford University Press, 1995).

130. John C. Burnham, "From Avant-Garde to Specialism: Psychoanalysis in America," *Journal of the History of the Behavioral Sciences* 15 (1979): 128–134; Hale, *The Rise and Crisis of Psychoanalysis*, Chapter 1.

131. John C. Burnham (ed.), *After Freud Left: A Century of Psychoanalysis in America* (Chicago: University of Chicago Press, 2012), 1.

132. Mark S. Micale, "The Psychiatric Body," in Roger Cooter and John Pickstone (eds.), *Companion to Medicine in the Twentieth Century* (London: Routledge, 2003), 323–346.

133. John Gach, "Culture & Complex: On the Early History of Psychoanalysis in America," in *Essays in the History of Psychiatry*, ed. Edwin R. Wallace IV and Lucius Pressley (Columbia, SC: William S. Hall Psychiatric Institute, 1980), 135–160.

134. Grob, *Mental Illness and American Society.*

135. Adolf Meyer, "The Dynamic Interpretation of Dementia Praecox," *American Journal of Psychology* 21 (1910): 385–403.

136. Gach, "Culture and Complex."

137. For more on Meyer, see Susan D. Lamb, *Pathologist of the Mind: Adolf Meyer and the Origins of American Psychiatry* (Baltimore, MD: Johns Hopkins University Press, 2014).

138. Hans Pols, "'Beyond the Clinical Frontiers.'"

139. Kuklick, "Philosophy at Yale," 321.

140. Eilhard von Domarus, "Prälogisches Denken in der Schizophrenie," *Zeitschrift für die gesamte Neurologie und Psychiatrie* 87, no. 1 (1923): 84–93; Eilhard von Domarus, "Zur Theorie des schizophrenen Denkens," *Zeitschrift für die Gesamte Neurologie und Psychiatrie* 108 (1927): 703–714; Eilhard von Domarus, "Über das Denken der Manischen und Depressiven," *Zeitschrift für die Gesamte Neurologie und Psychiatrie* 112 (1928): 632–635; Eilhard von Domarus, *Das Denken un seine krankhaften Störungen* (Leipzig: Kabitzsch, 1929). For more on disease categorization, see Richard Noll, *American Madness: The Rise and Fall of Dementia Praecox* (Cambridge, MA: Harvard University Press, 2011).

141. Gunter Walter Eilhard Alfred von Domarus, CV, Folder "Von Domarus, Eilhard," Series I, WSM Papers, APS. Between 1928 and 1930 von Domarus was not only a Research Fellow of the Rockefeller Foundation but also a Sterling Fellow at Yale, and taught psychiatry and mental hygiene at Yale from 1930 to 1932. Following this, Von Domarus began work at Rockland for three years.

142. Eilhard von Domarus, "The Logical Structure of Mind: An Inquiry into the Philosophical Foundation of Psychology and Psychiatry," in *Communication: Theory and Research. Proceedings of the First symposium,* ed. L. Thayer (Springfield IL: Charles C. Thomas, 1967), 351–411.

143. Von Domarus, "The Logical Structure of Mind," 353–354.

144. Von Domarus, "The Logical Structure of Mind," 364, 367.

145. Von Domarus, "The Logical Structure of Mind," 363.

146. Eilhard von Domarus, (1944). "The Specific Laws of Logic in Schizophrenia," in *Language and Thought in Schizophrenia: Collected Papers Presented at the Meeting of the American Psychiatric Association, May 12, 1939,* ed. J. S. Kasanin (Berkeley: University of California Press), 104–114.

147. Von Domarus, "The Specific Laws of Logic," 109–111; Robert Klee, "The Phrenetic Calculus: A Logician's View of Disordered Logical Thinking in Schizophrenia," *Behavior and Philosophy* 20/21, 2/1 (1993): 49–61.

148. Von Domarus to McCulloch, 29 February 1940, Folder "Von Domarus, Eilhard," Series I, WSM Papers, APS.

149. In 1948, McCulloch, apparently, was working on editing von Domarus's thesis and sent it in 1950 to a long list of people, including Percival Bailey, Frank Fremont-Smith, Heinrich Klüver, Jerry Lettvin, Margaret Mead, Claude Shannon, and Gerhardt von Bonin.

150. McCulloch himself had actually first met Northrop in fall 1923, while McCulloch was still a medical student at Columbia. Northrop taught mathematical biophysics to Frank Pike's graduate seminar on the theory of the nervous system, in which McCulloch was a participant (Filmer S. C. Northrop, "On W. S. McCulloch," in *The Collected Works of Warren S. McCulloch*, Volume I, ed. Rook McCulloch (Salinas, CA: Intersystems, 1989), 363–369). Their paths would cross again during the mid-1930s at Yale.

151. Bruce Kuklick, *The Rise of American Philosophy*, 566.

152. A. DiPalma, "Prof. F. S. C. Northrop Dies at 98; Saw Conflict Resolution in Science," *New York Times*, 23 July 1992, 89.

153. Filmer S. C. Northrop, "The History of Modern Physics in Its Bearing on Biology and Medicine," *Yale Journal of Biology and Medicine* 10 (1938): 209–232.

154. Filmer S. C. Northrop, "The Method and Theories of Physical Science and Their Bearing upon Biological Organization," *Growth (Supplement, Second Symposium on Development and Growth)*, 4, 127–154.

155. Northrop, "Method and Theories," 128.

156. Filmer S. C. Northrop, *Science and First Principles* (New York: Macmillan, 1931), xiii.

157. Northrop, "The History of Modern Physics," 231. While Northrop was clearly concerned with logic, philosophy of science, and epistemology, his views on logic and science were distinct from the logical positivists. Northrop shared the logical positivists' emphasis on the physical sciences as an ideal science, and viewed the sciences hierarchically. However, while the logical positivists viewed logic as a formal system, empty of empirical content, Northrop had a broad view of logic and a pluralist, relativist view of scientific method. Filmer S. C. Northrop, *Logic in the Sciences and Humanities* (New York: Macmillan, 1947).

158. Warren S. McCulloch, "Lekton, Being a Belated Introduction to the Thesis of Eilhard von Domarus," in *Communication: Theory and Research. Proceedings of the First Symposium*, ed. L. Thayer (Springfield, IL: Charles C. Thomas, 1967), 348–350; 350.

## Chapter 3

1. For the wider cultural context and implications of this development, see Rebecca Lemov, *World as Laboratory: Experiments with Mice, Mazes, and Men* (New York: Farrar, Straus, and Giroux, 2005). I borrow the term "interstitial" from Joel Isaac, who uses it to describe the group of Harvard academics interested in scientific philosophy, who operated between traditional established academic disciplines during the mid-twentieth century. See Joel Isaac, *Working Knowledge: Making the Human Sciences from Parsons to Kuhn* (Cambridge, MA: Harvard University Press, 2012), 22–27.

2. Susan Currell, "Eugenic Decline and Recovery in Self-Improvement Literature of the Thirties," in *Popular Eugenics: National Efficiency and American Mass Culture in the 1930s*, ed. Susan Currell and Christina Cogdell (Athens: Ohio University Press, 2006), 44–69.

3. Roderick Nash, *The Nervous Generation: American Thought, 1917–1930* (Chicago: Rand McNally, 1971), 52; Eugene Taylor, *Shadow Culture: Psychology and Spirituality in America* (Washington, DC: Counterpoint, 1999), Chapter 10.

4. Smith Ely Jelliffe to Ernest Jones, 10 February 1927, quoted in Gerald N. Grob, *The Inner World of American Psychiatry, 1890–1940: Selected Correspondence* (New Brunswick, NJ: Rutgers University Press, 1985), 224–225.

5. Paul V. Murphy, *The New Era: American Thought and Culture in the 1920s* (Lanham, MD: Rowman & Littlefield, 2012), 21–22.

6. Currell, "Eugenic Decline," 54.

7. John C. Burnham, *How Superstition Won and Science Lost: Popularizing Science and Health in the United States* (New Brunswick and London: Rutgers University Press, 1987), 97.

8. Jack D. Pressman, *Last Resort: Psychosurgery and the Limits of Medicine* (Cambridge: Cambridge University Press, 1998), 27.

9. Elizabeth Lunbeck, *The Psychiatric Persuasion: Knowledge, Gender, and Power in Modern America* (Princeton, NJ: Princeton University Press, 1994), Chapter 3.

10. Susan D. Lamb, *Pathologist of the Mind: Adolf Meyer and the Origins of American Psychiatry* (Baltimore, MD: Johns Hopkins University Press, 2014).

11. Adolf Meyer, *Psychobiology: A Science of Man*, compiled and edited by Eunice E. Winters and Anna Mae Bowers (Springfield, IL: Charles C. Thomas, 1957), xiii, 44, 48.

12. Adolf Meyer to Abraham Flexner, 20 April 1927, cited in Grob, *The Inner World of American Psychiatry*, 175–181.

13. Meyer, *Psychobiology*, 6.

14. Madison Bentley and E. V. Cowdry (eds.), *The Problem of Mental Disorder: A Study Undertaken by the Committee on Psychiatric Investigations, National Research Council* (New York and London: McGraw-Hill, 1934), vi.

15. Ibid.

16. Ibid., vii.

17. Meyer contributed to the volume yet was unhappy that the editors "sought salvation in more or less artificial synthesis. ..." Meyer to John C. Whitehorn, 12 February 1932, in Grob, *The Inner World of American Psychiatry*, 219.

18. Grob, *The Inner World of American Psychiatry*, 230.

19. Commission on Medical Education, *Final Report of the Commission on Medical Education* (New York: Office of the Director of Study, 1932), 213–214. For more on psychiatry and neurology in the context of medical education, see J. Brand, "Neurology and Psychiatry," in *The Education of American Physicians: Historical Essays*, ed. Ronald L. Numbers (Berkeley: University of California Press, 1980), 226–249.

20. Pressman, *Last Resort*. The emergence and flourishing of the practice of lobotomy in this context owed much to this framework, as desperation for cures fuelled the development of biological psychiatry.

21. Mark Dowie, *American Foundations: An Investigative History* (Cambridge, MA: MIT Press, 2001), 49.

22. Dowie, *American Foundations*, 51.

23. Olivier Zunz, *Philanthropy in America: A History* (Princeton and Oxford: Princeton University Press, 2012). For more on the impact of philanthropy on public policy, see Judith Sealander, *Private Wealth and Public Life: Foundation Philanthropy and the Reshaping of American Social Policy from the Progressive Era to the New Deal* (Baltimore and London: Johns Hopkins University Press, 1997).

24. Lily E. Kay, *The Molecular Vision of Life: Caltech, the Rockefeller Foundation, and the Rise of the New Biology* (New York: Oxford University Press, 1993), 8.

25. Kay, *Molecular Vision of Life*, 31.

26. Robert E. Kohler, *Partners in Science: Foundations and Natural Sciences 1900–1945* (Chicago: University of Chicago Press, 1991), 233.

27. "The Medical Sciences" (April 1933), Folder "Excerpt, History, 'The Medical Sciences'", RG 3, Series 900, Box 22.168, pp. 10–25, p. 19, Rockefeller Archive Center, Pocantico Hills, New York; hereafter RAC.

28. Kay, *Molecular Vision of Life*; Kohler, *Partners in Science*.

29. Rockefeller Foundation Trustees Bulletin Number 120, April 1 1950, RG 3, Series 915, Box 1.2, RAC. See also Raymond B. Fosdick, *The Story of the Rockefeller Foundation* (New York: Harper, 1952), Chapter X.

30. See Walter Bromberg, *Psychiatry Between the Wars, 1918–1945: A Recollection* (Westport, CT: Greenwood Press, 1982).

31. More on Gregg's role in Foundation medical research policy can be found in Brown, "Alan Gregg"; Jack D. Pressman, "Human Understanding: Psychosomatic Medicine and the Mission of the Rockefeller Foundation," in *Greater than the Parts: Holism in Biomedicine, 1920–1950*, ed. Christopher Lawrence and George Weisz (New York and Oxford: Oxford University Press, 1998), 189–208; and William H. Schneider, "The Men Who Followed Flexner: Richard Pearce, Alan Gregg, and the Rockefeller Foundation Medical Divisions, 1919–1951," in *Rockefeller Philanthropy and Modern Biomedicine: International Initiatives from World War I to the Cold War*, ed. William H. Schneider (Bloomington: Indiana University Press, 2002), 7–60.

32. David L. Edsall, "Memorandum Regarding Possible Psychiatric Developments," 3 October 1930, RG 3, Series 906, Box 2.19, RAC.

33. Daniel Todes, *Pavlov's Physiology Factory: Experiment, Interpretation, Laboratory Enterprise* (Baltimore and London: Johns Hopkins University Press, 2002); Kenton Kroker, *The Sleep of Others and the Transformations of Sleep Research* (University of Toronto Press, 2007), 215–218.

34. Edsall, "Memorandum," 5.

35. Edsall, "Memorandum," 5.

36. Pressman, "Human Understanding," 193–194.

37. "Proposed Future Program" (General Statement), April 1933, RG 3, Series 900, Box 22.168, p. 63, RAC.

38. "The Medical and Natural Sciences," 13 December 1933, RG 3, Series 900, Box 22.168, RAC.

39. Pressman, "Human Understanding," 193.

40. Alan Gregg, "The Strategy of Our Program in Psychiatry" from *Confidential Monthly Report for the Information of the Trustees*, Nov. 1, 1937, RG 3, Series 906, Box 1.4, RAC. Grant Action, Northwestern Medical School—Institute of Neurology, 15 Feb. 1935, RG 1.1 Series 216A Box 5.52, RAC.

41. Louise Marshall, "Instruments, Techniques, and Social Units in American Neurophysiology, 1870–1950," in *Physiology in the American Context: 1850–1940*, ed. Gerald L. Geison (Bethesda, MD: American Physiological Society, 1987), 351–369; 358.

42. Borell, "Instrumentation," 58.

43. Marshall, "Instruments, Techniques, and Social Units."

44. Cornelius Borck, *Hirnströme: Eine Kulturgeschichte Der Elektroencephalographie* (Wallstein Verlag, 2005); Cornelius Borck, "Between Local Cultures and National Styles: Units of Analysis in the History of Electroencephalography," *Comptes Rendus Biologies* 329 (2006): 450–459.

45. Adele E. Clarke and Joan H. Fujimura, "What Tools? Which Jobs? Why Right?" in *The Right Tools for the Job: At Work in Twentieth-Century Life Sciences*, ed. Adele E. Clarke and Johan Fujimura (Princeton, NJ, and Oxford: Princeton University Press, 1992), 3–44.

46. The classic historical account of the nineteenth-century origins of cerebral localization is Robert M. Young, *Mind, Brain and Adaptation in the Nineteenth Century: Cerebral Localization and Its Biological Context from Gall to Ferrier* (New York: Oxford University Press, 1970). For more on the American context, see Tara H. Abraham, "Transcending Disciplines: Scientific Styles in Studies of the Brain in Mid-Twentieth Century America," *Studies in History and Philosophy of Biological and Biomedical Sciences* 43 (2012): 552–568.

47. Star, *Regions of the Mind*, 7.

48. Susan Leigh Star, "Triangulating Basic and Clinical Research: British Localizationists, 1870–1906," *History of Science* 24 (1986): 29–48, Star; *Regions of the Mind*, Chapter 4.

49. Star, *Regions of the Mind*, 159–160.

50. Gerald L. Geison, "Divided We Stand: Physiologists and Clinicians in the American Context," in *The Therapeutic Revolution: Essays in the Social History of American Medicine*, ed. Morris J. Vogel and Charles E. Rosenberg (Philadelphia: University of Pennsylvania Press, 1979), 67–90.

51. Charles S. Sherrington, *The Integrative Action of the Nervous System* (New York: Charles Scribner's Sons, 1906), 2.

52. Star, *Regions of the Mind*, 5.

53. Karl Lashley, "Studies of Cerebral Function in Learning VII: The Relation between Cerebral Mass, Learning, and Retention," *Journal of Comparative Neurology* 41 (1926): 1–58; Nadine M. Weidman, *Constructing Scientific Psychology: Karl Lashley's Mind-Brain Debates* (Cambridge University Press, 1999), 67.

54. This list includes Wilder Penfield, Harvey Cushing, Derek Denny-Brown, John C. Eccles, and F. M. R. Walshe. Roger Smith, "The Embodiment of Value: C. S. Sherrington and the Cultivation of Science," *British Journal for the History of Science* 33 (2000): 283–311, 284.

55. Roger Smith, "Representations of Mind: C. S. Sherrington and Scientific Opinion, ca. 1930–1950," *Science in Context* 14, no. 4 (2001): 511–539, 517. See also Pressman, *Last Resort*, Chapter 2.

56. Gerard N. Burrow, *A History of Yale's School of Medicine: Passing Torches to Others* (New Haven and London: Yale University Press, 2002), 128.

57. Pressman, *Last Resort*, 47.

58. Mark A. May, "A Retrospective View of the Institute of Human Relations at Yale," *Behavior Science Notes* 6 (1971): 141–172.

59. Jill G. Morawski,"Organizing Knowledge and Behavior at Yale's Institute of Human Relations," *Isis* 77 (1986): 219–242.

60. May, "A Retrospective View," 143.

61. Magoun, *American Neuroscience*, 83.

62. H. E. Hoff, "John Fulton's Contribution to Neurophysiology," *Journal of the History of Medicine and Allied Sciences* 17 (1962): 16–71; Magoun, *American Neuroscience*, 87.

63. Fulton, "Postgraduate Training in Psychiatry—Outline," June 1933, RG 1.1, Series 200A, Box 120.1480, RAC.

64. Fulton to Gregg, 25 February 1933, RG 1.1, Series 200A, Box 120.1480, RAC.

65. Fulton was referring to Rudolf Magnus, who will be discussed below.

66. John F. Fulton, "Yale University—Neurophysiology, Memorandum Concerning Postgraduate Instruction in Neurophysiology," 12 June 1933, RG 1.1, Series 200A, Box 120.1482, RAC.

67. Grant Action, Yale University—Neurophysiology, 9 January 1934, RG 1.1, Series 200A, Box 120.1480, RAC. The grant was renewed in 1935 and 1936.

68. John F. Fulton, "Yale University School of Medicine, Department of Physiology," in *Methods and Problems of Medical Education. No. 20*, pp. 17–23 (New York: Division of Medical Education, Rockefeller Foundation, 1932).

69. John F. Fulton and Ralph W. Gerard, "J. G. Dusser de Barenne 1885–1940," *Journal of Neurophysiology* 3 (1940): 283–292.

70. E.g., Johannes G. Dusser de Barenne, "Die Strychninwirkung auf das Zentralnervensystem. I. Die Wirkung des Strychnins auf die Reflextätigkeit der Intervertebralganglia," *Folia Neuro-Biologica Leipzig* 4 (1910): 467–474. Sherrington had used strychnine in reflex studies as well (1906) where he observed that the substance caused skeletal muscles to contract.

71. For an analysis of the history of strychnine in chemical, physiological, and pharmacological contexts, see Jonathan Simon, "Naming and Toxicity: A History of Strychnine," *Studies in History and Philosophy of Biological and Biomedical Sciences* 30, no. 4 (1999): 505–525.

72. For more on Magnus, see Otto Magnus, *Rudolf Magnus, Physiologist and Pharmacologist: A Biography* (Amsterdam: Royal Netherlands Academy of Arts and Sciences, 2002).

73. Rudolf Magnus, *Körperstellung*. (Berlin: Julius Springer, 1924). In 1927 Magnus was nominated for a Nobel Prize along with de Kleijn, although he passed away before the decision had been made.

74. Rudolf Magnus, *Goethe als Naturforscher* (1906); Trans. Heinz Norden as *Goethe as a Scientist* (New York: Henry Schuman, 1950).

75. Magnus was to have delivered five lectures in 1928. He completed—but did not deliver—three lectures before his untimely death while on holiday in July 1927. Rudolf Magnus, "The Physiological *a Priori*," *Lane Lectures on Experimental Pharmacology and Medicine*, Stanford University Publications, University Series, Medical Sciences. Volume II, No. 3. (Stanford CA: Stanford University Press, 1930), 97–103.

76. Magnus, "The Physiological *a Priori*," 103.

77. Johannes G. Dusser de Barenne, "Experimental Researches on Sensory Localisations in the Cerebral Cortex," *Quarterly Journal of Experimental Physiology* 9 (1916): 355–390, 355.

78. Dusser de Barenne, "Experimental Researches on Sensory Localisations in the Cerebral Cortex," 358.

79. Johannes G. Dusser de Barenne, "Experimental Researches on Sensory Localization in the Cerebral Cortex of the Monkey (Macacus)," in *Proceedings of the Royal Society of London, Series B* 96, no. 676 (1924): 272–291.

80. Fulton and Gerard, "J. G. Dusser de Barenne 1885–1940," 285.

81. Records of the Dean, 1888–1960, School of Medicine, Boxes 79 and 88, Yale University Archives.

82. This was not cortical *inhibition,* which was cessation of a response to stimulation if a second, "antagonistic" part of the cortex was simultaneously stimulated. Rather, extinction described the phenomenon of decreased response to stimulation of the cortex. Johannes G. Dusser de Barenne and Warren S. McCulloch, "An 'Extinction' Phenomenon on Stimulation of the Cerebral Cortex," *Proceedings of the Society for Experimental Biology and Medicine* 32 (1934): 524–527.

83. Johannes G. Dusser de Barenne and Warren S. McCulloch, "Functional Boundaries in the Sensori-motor Cortex of the Monkey," *Proceedings of the Society for Experimental Biology and Medicine* 35 (1936): 329–331.

84. Johannes G. Dusser de Barenne and Warren S. McCulloch, "Some Effects of Local Strychninization on Action Potentials of the Cerebral Cortex of the Monkey," *Transactions of the American Neurological Association (62nd Annual Meeting, Atlantic City, NJ, June 1–3, 1936)*, 171.

85. Johannes G. Dusser de Barenne and Warren S. McCulloch, "Functional Organization in the Sensory Cortex of the Monkey (Macaca Mulatta)," *Journal of Neurophysiology* 1 (1938): 69–85; Johannes G. Dusser de Barenne and Warren S. McCulloch, "The Direct Functional Interrelation of Sensory Cortex and Optic Thalamus," *Journal of Neurophysiology* 1 (1938): 176–186; Johannes G. Dusser de Barenne and Warren S. McCulloch, "Sensorimotor Cortex, Nucleus Caudatus, and Thalamus Opticus," *Journal of Neurophysiology* 1 (1938): 364–376; Johannes G. Dusser de Barenne, Warren S. McCulloch, and Teizo Ogawa, "Functional Organization in the Face-Subdivision of the Sensory Cortex of the Monkey (Macaca Mulatta)," *Journal of Neurophysiology* 1 (1938): 436–441.

86. Johannes G. Dusser de Barenne, "Some Aspects of the Problem of "Corticalization" of Function and of Functional Localization in the Cerebral Cortex," in *Localization of Function in the Cerebral Cortex. In Proceedings of the Association for Research in Nervous and Mental Disease, Vol. XIII*, ed. Samuel T. Orton, John F. Fulton, and Samuel K. Davis (Baltimore, MD: Williams and Wilkins, 1934), 85–106; 96.

87. Dusser de Barenne, "Some Aspects of the Problem of 'Corticalization,'" 103.

88. Johannes G. Dusser de Barenne and Warren S. McCulloch, "Physiological Delimitation of Neurones in the Central Nervous System," *American Journal of Physiology* 127 (1939): 620–628. This was a departure from anatomical neuronography, which simply involved staining cell bodies in the cortex and physically distinguishing parts of the cortex from one another. Jerome Y. Lettvin, "Strychnine Neuronography," in *The Collected Works of Warren S. McCulloch*, Volume I, ed. Rook McCulloch (Salinas, CA: Intersystems., 1989), 50–58; 50.

89. Fulton and Gerard, "Johannes Gregorius Dusser de Barenne," 286.

90. In the words of Fulton and Gerard, "[t]hese were men of action, men of experiment, who never allowed their deductions to usurp the place of experiment." Fulton and Gerard, "Johannes Gregorius Dusser de Barenne," 286.

91. Warren S. McCulloch, "Johannes Gregorius Dusser de Barenne (1885–1940)," *Yale Journal of Biology and Medicine* 12 (1940): 743–746.

92. Dusser de Barenne also played a role in the founding of the *Journal of Neurophysiology* in 1938, and was an active editor of the journal until his death in 1940.

93. Interview with Taffy Holland and David McCulloch, 13 July 2002, Old Lyme, Connecticut.

94. Mary Jean Vasiloff, Personal Communication, 27 January 2015.

95. McCulloch, "Recollections," 33.

96. Percival Bailey, Johannes G. Dusser de Barenne, Hugh W. Garol, and Warren S. McCulloch, "Sensory Cortex of Chimpanzee," *Journal of Neurophysiology* 3 (1940): 469–485.

97. Laurence D. Smith, *Behaviorism and Logical Positivism: A Reassessment of the Alliance* (Stanford: Stanford University Press, 1986); Francesca Bordogna, *William James at the Boundaries: Philosophy, Science, and the Geography of Knowledge* (Chicago: University of Chicago Press, 2008); Andrew Jewett, *Science, Democracy, and the American University: From the Civil War to the Cold War* (Cambridge University Press, 2012); Joel Isaac, *Working Knowledge*; Gary L. Hardcastle and Alan W. Richardson (eds.), *Logical Empiricism in North America, Minnesota Studies in the Philosophy of Science, Volume XVIII*, (Minneapolis, London: University of Minnesota Press, 2003).

98. See, especially, Alan W. Richardson and Gary L. Hardcastle, "Introduction: Logical Empiricism in North America," in *Logical Empiricism in North America, Minnesota Studies in the Philosophy of Science, Volume XVIII*, ed. Gary L. Hardcastle and Alan W. Richardson (Minneapolis, London: University of Minnesota Press, 2003), vii–xxix; Alan Richardson, "Scientific Philosophy as a Topic for History of Science," *Isis* 99, no. 1 (March 2008): 88–96.

99. See Rudolf Carnap, *Der Logische Aufbau der Welt*; Alan Richardson, "Scientific Philosophy."

100. Alan W. Richardson, "Logical Empiricism, American Pragmatism, and the Fate of Scientific Philosophy in North America," in *Logical Empiricism in North America, Minnesota Studies in the Philosophy of Science, Volume XVIII*, ed. Gary L. Hardcastle and Alan W. Richardson (Minneapolis, London: University of Minnesota Press, 2003), 1–24.

101. Richardson, "Scientific Philosophy," 90.

102. Smith, *Behaviorism and Logical Positivism*, 18.

103. Smith, *Behaviorism and Logical Positivism*, 154–182.

104. Kenneth Spence, "Foreword," in Hull, *Principles of Behavior*, xv.

105. Hull, *Principles of Behavior*, Chapter 1.

106. Hull, *Principles of Behavior*, 14.

107. May, "A Retrospective View," 160.

108. Weidman, *Constructing a Scientific Psychology*, 123.

109. Smith, *Behaviorism and Logical Positivism*, 158–159; Keith Gunderson, "Cybernetics," *The Encyclopedia of Philosophy* (New York and London: Macmillan, 1967), 280–284; Rebecca Lemov, *World as a Laboratory*, 85.

110. Smith, *Behaviorism and Logical Positivism*, 152; Clark L. Hull, "Mind, Mechanism, and Adaptive Behavior," *Psychological Review* 44, no. 1 (1937): 1–32.

111. Hull, "Mind, Mechanism, and Adaptive Behavior," 5.

112. Smith, *Behaviorism and Logical Positivism*, 180–81.

113. May, "A Retrospective View."

114. Morawski, "Organizing Knowledge and Behavior," 234–235.

115. Clark L. Hull, *Principles of Behavior: An Introduction to Behavior Theory* (New York: Appleton-Century-Crofts, 1943), vii.

116. Morawski, "Organizing Knowledge and Behavior," 240–241.

117. Hull and Mowrer, *Hull's Psychological Seminars, 1936–1938*, 3.

118. May, "A Retrospective View," 162.

119. Clark L. Hull and C. H. Mowrer, *Hull's Psychological Seminars, 1936–1938*, Notices and Abstracts of Proceedings, Institute of Human Relations, YUA.

120. Clark L. Hull, *Psychology Seminar Memoranda*, Yale University.

121. Hull and Mowrer, *Hull's Psychological Seminars, 1936–1938*, 23, 29.

122. Northrop, "On Warren S. McCulloch," in *The Collected Works of Warren S. McCulloch*, Volume 1, ed. Rook McCulloch (Salinas, CA: INtersystems, 1989), 367. Northrop viewed his own contributions at these seminars as the reason for his regular participation in the Macy Meetings on Cybernetics.

123. This view on logic owes less to the tradition of logical positivism and more to American pragmatism, e.g., John Dewey.

124. E.g., Clark L. Hull, *Mathematico-Deductive Theory of Rote Learning* (New Haven: CT: Yale University Press, 1940).

125. Anonymous, "Huxley, the Brain, and the Mind: Professor Adrian's Huxley Lecture," *British Medical Journal* (27 Nov 1937): 1082–1083, 1082.

126. Star, *Regions of the Mind*.

127. McCulloch to Fulton, 20 January 1940, and 4 February 1940, Folder Fulton, John F., Series I, WSM Papers, APS.

128. McCulloch wrote to Bailey for feedback on the manuscript. McCulloch to Bailey, 21 June 1940, Folder "Bailey, Percival II," McCulloch Papers, Series I, APSL. Von Bonin was apparently planning to come to Dusser de Barenne's lab at Yale that

fall. Bailey and McCulloch were quite close (see Bailey to McCulloch, 25 December 1940, Folder "Bailey, Percival II," McCulloch Papers, Series I, APSL).

129. McCulloch to Bailey, 1 February 1941, Folder "Bailey, Percival II," McCulloch Papers, Series I, APSL. McCulloch feared losing his position at Yale and his staff and felt the need to look elsewhere, hinting that he'd like to come to Illinois.

130. McCulloch to Bailey, 4 February 1941, Folder "Bailey, Percival II," McCulloch Papers, Series I, APSL. Ward's Island in Manhattan is the home of several psychiatric hospitals.

131. Fulton to McCulloch, 5 March 1941, Folder "Fulton, John F.," McCulloch Papers, Series I, APSL. Indeed, Fulton reported to Gregg in February 1941 that the Board of Permanent Officers at the medical school had just voted to have the Laboratory of Neurophysiology, under McCulloch's direction since Dusser de Barenne's death, to be incorporated into the Department of Physiology (Excerpt from Fulton to Gregg, 4 February 1941, RG 1.1, Series 200A, Box 120.1481, RAC).

132. J. R. Hughes, "Development of Clinical Psychology in the Chicago Area," *Journal of Clinical Neurophysiology* 10, no. 3 (1993), 298–303. During the 1930s, Bailey had organized an informal "neurology club" in Chicago (Marshall, "Instruments, Techniques, and Social Units"). He had hoped to establish a division of neurosciences at the University of Chicago, but when the administration proved unsupportive, Bailey moved to the University of Illinois, and became the director of the newly formed Neuropsychiatric Institute in 1939 (Bonnie Ellen Blustein, "Percival Bailey and Neurology at the University of Chicago, 1928–1939," *Bulletin of the History of Medicine* 66 (1992): 90–113.).

## Chapter 4

1. Warren S. McCulloch and Walter Pitts, "A Logical Calculus of the Ideas Immanent in Nervous Activity," *Bulletin of Mathematical Biophysics* 5 (1943): 115–133.

2. E.g., Lily E. Kay, "From Logical Neurons to Poetic Embodiments of Mind: Warren S. McCulloch's Project in Neuroscience," *Science in Context* 14, no. 4 (2001): 591–614; Michael A. Arbib, "Warren McCulloch's Search for the Logic of the Nervous System," *Perspectives in Biology and Medicine*, 43, no. 2 (2000): 193–216; Steve J. Heims, *Constructing a Social Science for Postwar America: The Cybernetics Group* (Cambridge, MA: MIT Press, 1993), Chapter 3; Elizabeth A. Wilson, *Affect and Artificial Intelligence* (Seattle and London: University of Washington Press, 2010); Jean-Pierre Dupuy, *The Mechanization of the Mind: On the Origins of Cognitive Science*, trans. M. B. DeBevoise (Princeton, N.J.: Princeton University Press, 2000); Gualtiero Piccinini, "The First Computational Theory of Mind and Brain: A Close Look at McCulloch and Pitts's 'Logical Calculus of Ideas Immanent in Nervous Activity'," *Synthese* 141, no. 2 (2004): 175–215; Kenneth Aizawa, "Warren McCulloch's Turn to Cybernetics: What

Walter Pitts Contributed," *Interdisciplinary Science Reviews* 37, no. 3 (September 2012): 206–217.

3. Bonnie Ellen Blustein, "Percival Bailey and Neurology at the University of Chicago, 1928–1939," *Bulletin of the History of Medicine* 66 (1992): 90–113.

4. D. J. Davis to Alan Gregg, 1 March 1935, RG 1.1, Series 200A, Box 112.1373, RAC. The University of Illinois medical school, the College of Physicians and Surgeons, was organized in 1882, and eventually was affiliated with Cook County Hospital for its clinical facilities.

5. Adolf Meyer, "H. Douglas Singer, M.D. 1875–1940," *Archives of Neurology and Psychiatry* 45, no. 1 (1941): 162–164.

6. H. Douglas Singer, "Undergraduate Teaching of Psychiatry, University of Illinois College of Medicine," 1935, RG 1.1, Series 200A, Box 112.1373, RAC.

7. Alan Gregg Interview with Dean Davis and H. Douglas Singer, 13 January 1936, RG 1.1, Series 200A, Box 112.1373, RAC.

8. Grant Action, University of Illinois—College of Medicine—Development of Psychiatry, 15 May 1936, RG 1.1, Series 200A, Box 112.1373, RAC.

9. Elizabeth Lunbeck, *The Psychiatric Persuasion: Knowledge, Gender, and Power in Modern America* (Princeton University Press, 1994), 46–47.

10. Pamphlet, The Illinois Neuropsychiatric Institute and the University of Illinois, Chicago, Illinois, 6 June 1942, Folder "Neuro-Psychiatric Institute," Box 1, "The Illinois Neuropsychiatric Institute and the University of Illinois," Series RS 53/38/815, University of Illinois at Urbana-Champaign Archives, Urbana, IL (hereafter UIUCA).

11. Grant Action, University of Illinois—Institute of Neurology, 15 November 1940, RG 1.1, Series 200A, Box 111.1371, RAC.

12. Memo, Board Meeting, 10 November 1937, "Expansion of Neurology and Neurological Surgery," Folder "Neuro-Psychiatric Institute," Box 54, Arthur Cutts Willard Papers, Series RS 2/9/1 (General Correspondence), UIUCA.

13. "The Illinois Neuropsychiatric Institute," excerpt from Trustees Confidential Bulletin, February 1949, RG 1.1, Series 200A, Box 112.1376, RAC.

14. Sveinbjorn Johnson to A. C. Willard, 12 May 1939, Folder "Neuro-Psychiatric Institute," Box 54, Arthur Cutts Willard Papers, Series RS 2/9/1 (General Correspondence), UIUCA.

15. John R. Hughes, "Development of Clinical Neurophysiology in the Chicago Area," *Journal of Clinical Neurophysiology* 10, no. 3 (1993): 298–303; 299.

16. Alan Gregg Diary, 18 October 1940, RG 1.1, Series 200A, Box 111.1371, RAC; "Form No. 2, Proposed Amendment to Section 5, Article II of Original Agreement

Between the Department of Public Welfare and the Board of Trustees of the University of Illinois" Folder "Neuro-psychiatric Institute," Box 54, Arthur Cutts Willard Papers, Series 9/2/1 (General Correspondence), UIUCA.

17. Alan Gregg Diary, 18 October 1940, RG 1.1, Series 200A, Box 111.1371, RAC.

18. "Illinois Neuropsychiatric Institute Made Welfare History—Address by Governor Dwight H. Green," *The Welfare Bulletin* 33, no. 8 (August 1942), 5, Box 1, Series RS 53/38/815, UIUCA.

19. Ibid., 6.

20. Francis J. Gerty to Alan Gregg, 11 October 1944, RG 1.1 Series 200A, Box 112.1374, RAC.

21. McCulloch's yearly salary of $5,000 was paid out of the University funding. Francis J. Gerty, "Prospectus of the Research Laboratory of the Illinois Neuropsychiatric Institute," ca. 1942/43, Warren S. McCulloch Papers, Folder "Gerty, Francis J./ III," Series I, WSM Papers, APS.

22. Warren S. McCulloch to Dorothea Dusser de Barenne, 1 December 1941, Folder "Dusser de Barenne, Dorothea," Series I, WSM Papers, APS.

23. Ibid.

24. Magoun, *American Neuroscience*, 100. He also contributed a review to Fulton's *Physiology of the Nervous System*. See John F. Fulton to Warren S. McCulloch, 26 February 1943, Folder "Fulton, John F./III," Series I, WSM Papers, APS.

25. Horace W. Magoun, "The Role of Research Institutes in the Advancement of Neuroscience: Ranson's Institute of Neurology," in Frederic G. Worden, Judith P. Swazey, and George Adelman (eds.), *The Neurosciences: Paths of Discovery* (Cambridge, MA: MIT Press, 1975), 515–527. Magoun had arrived at the INI in 1943.

26. Pamphlet, Illinois Neuropsychiatric Institute and the University of Illinois, 6 June 1942, Box 1, "The Illinois Neuropsychiatric Institute and the University of Illinois," Series RS 53/38/815, UIUCA.

27. Francis J. Gerty, "Prospectus of the Research Laboratory of the Illinois Neuropsychiatric Institute."

28. Jerome Lettvin, "Warren as a Teacher," in *The Collected Works of Warren S. McCulloch, Volume I*, ed. Rook McCulloch (Salinas, CA: Intersystems), 332–338; 337; David McCulloch, Personal Communication, 4 February 2015.

29. For more on Lettvin, see Jerome Y. Lettvin, "Jerome Y. Lettvin," in Larry R. Squire (ed.), *The History of Neuroscience in Autobiography, Volume 2* (San Diego: Academic Press, 1998), 224–242; Elizabeth A. Wilson, *Affect and Artificial Intelligence*, Chapter 4; Tara H. Abraham, "'The Materials of Science, the Ideas of Science, and

the Poetry of Science': Warren McCulloch and Jerry Lettvin," *Interdisciplinary Science Reviews* 37, no. 3 (September 2012): 269–86.

30. Jerome Y. Lettvin, "Jerome Lettvin," 227.

31. See, e.g., Lettvin, "Jerome Y. Lettvin"; Neil R. Smalheiser, "Walter Pitts," *Perspectives in Biology and Medicine* 43, no. 2 (2000): 217–226; Abraham, "(Physio)logical Circuits"; Elizabeth A. Wilson, *Affect and Artificial Intelligence*, Chapter 4; Flo Conway and Jim Siegelman, *Dark Hero of the Information Age: In Search of Norbert Wiener, The Father of Cybernetics* (New York: Basic Books, 2005).

32. Jerome Y. Lettvin, "Interview with James A. Anderson and E. Rosenfeld," in *Talking Nets: An Oral History of Neural Networks* (Cambridge, MA: MIT Press, 1998), 2.

33. I treat Rashevsky's story in detail in Tara H. Abraham, "Nicolas Rashevsky's Mathematical Biophysics," *Journal of the History of Biology* 37 (2004): 333–385. See also Maya M. Shmailov, "Nicolas Rashevsky's Pencil-and-Paper Biology," in *Outsider Scientists: Routes to Innovation in Biology*, ed. Oren Harman and Michael R. Dietrich (Chicago: University of Chicago Press, 2013), 161–180; and Evelyn Fox Keller, *Making Sense of Life: Explaining Biological Development with Models, Metaphors, and Machines* (Cambridge, MA: Harvard University Press, 2002).

34. E.g., Nicolas Rashevsky, "Possible Brain Mechanisms and Their Physical Models," *Journal of General Psychology*, 5 (1931): 368–406; Nicolas Rashevsky, "On the Theory of Nerve Conduction," *Journal of General Physiology* 14 (1931): 517–528; Nicolas Rashevsky, "Outline of a Physico-Mathematical Theory of Excitation and Inhibition," *Protoplasma* 20 (1933): 42–56.

35. E.g., Archibald V. Hill, "A New Mathematical Treatment of Changes of Ionic Concentration in Muscle and Nerve under the Action of Electric Currents, with a Theory as to Their Mode of Excitation," *Journal of Physiology* 40 (1910): 190–224.

36. For later developments on this front, see Phillip R. Sloan, "Molecularizing Chicago—1945–1965: The Rise, Fall, and Rebirth of the University of Chicago's Biophysics Program," *Historical Studies in the Natural Sciences* 44, no. 4 (September 2014): 364–412.

37. Alfred Lotka, *Elements of Physical Biology* (Baltimore, MD: Williams and Wilkins, 1925).

38. Vito Volterra, *Leçons sur la Théorie Mathématique de la Lutte pour la Vie* (Paris: Gauthier-Villars et cie, 1931).

39. Nicolas Rashevsky, *Mathematical Biophysics: Physicomathematical Foundations of Biology* (Chicago: University of Chicago Press, 1938), vii.

40. Nicolas Rashevsky, *Mathematical Biophysics*, 1, emphasis in original.

41. Warren Weaver Interview, January 19, 1939, RG 1.1, Series 216D, Box 11.148, RAC.

42. A. F. Bartholomay, George Karreman, and H. D. Landahl, "Nicolas Rashevsky," *Bulletin of Mathematical Biophysics* 34 (1972) (no pagination).

43. For example, in 1939, E. R. Hilgard and C. P. Stone examined neurophysiological theories of conditioning, and dubbed Rashevsky's work on the subject "alternative speculations which suffered from a lack of confirmatory evidence." E. R. Hilgard and C. P. Stone, "Physiological Psychology," *Annual Review of Physiology* 1 (1939): 471–487, p. 478.

44. Shmailov, "Nicolas Rashevsky's Pencil-and-Paper Biology."

45. Cowan, "Interview with James A. Anderson and Edward Rosenfeld," 104–105.

46. "Mailing list for seminar notices," Folder Mathematical Biophysics Seminar, Box 3, Nicolas Rashevsky Papers, University of Chicago Department of Special Collections (hereafter UCDSC). I thank Maya Shmailov for drawing my attention to this document.

47. McCulloch himself reported that in 1941 he had presented some of his ideas on neural activity to Rashevsky's seminar, and had met Pitts here. Warren S. McCulloch, "Recollections of the Many Sources of Cybernetics," in *The Collected Works of Warren S. McCulloch, Volume 1*, ed. Rook McCulloch (Salinas, CA: Intersystems, 1989), 21–49.

48. Lettvin, "Jerome Y. Lettvin," 227.

49. Conway and Siegelman, *Dark Hero*, 139.

50. Rook McCulloch, "Foreword," in *The Collected Works of Warren S. McCulloch, Volume 1*, ed. Rook McCulloch (Salinas, CA: Intersystems, 1989), 2.

51. Raisbeck in Conway and Siegelman, *Dark Hero*, 202.

52. Personal communication with Taffy Holland, McCulloch's daughter, 13 July 2002.

53. Abraham, "The Materials of Science."

54. Letter to participants from Frank Fremont-Smith, 11 May 1942, Folder "Fremont Smith, Frank/I," Series I, McCulloch Papers, APSL. More on the Macy Foundation agenda can be found in Heims, *The Cybernetics Group*, Chapter 7.

55. Letter to participants from Frank Fremont-Smith, 11 May 1942.

56. For the context of this work, see Peter Galison, "The Ontology of the Enemy: Norbert Wiener and the Cybernetic Vision," *Critical Inquiry* 21, no.1 (Autumn 1994): 228–266.

57. Arturo Rosenblueth, Norbert Wiener, and Julian Bigelow, "Behavior, Purpose, and Teleology," *Philosophy of Science* 10 (1943): 18–24, 24.

58. McCulloch to Fremont-Smith, 24 June 1942, Folder "Fremont-Smith/I," Series I, McCulloch Papers, APSL.

59. McCulloch to Fremont-Smith, 24 June 1942.

60. McCulloch to Fremont-Smith, 24 June 1942. McCulloch addresses the letter to "Doctor Fremont-Smith," indicating formality; thus McCulloch likely did not have a relationship with Fremont-Smith for long prior to 1942.

61. Bertrand Russell, *Introduction to Mathematical Philosophy* (London: George Allen and Unwin, 1920), 169.

62. Paul Schrecker, "Leibniz and the Art of Inventing Algorisms," *Journal of the History of Ideas* 8 (1947): 107–116.

63. Jerome Y. Lettvin, "Interview with James A. Anderson and Edward Rosenfeld," 3.

64. Alan M. Turing, "On Computable Numbers, with an Application to the *Entscheidungsproblem*," *Proceedings of the London Mathematical Society, Series 2,* 42 (1936–7): 230–265.

65. Smalheiser, "Walter Pitts," 219.

66. Interview with Jerome Y. Lettvin by Flo Conway and Jim Siegelman, 12 December 1999, in *Dark Hero,* 139. See also Lettvin, *Talking Nets,* 3.

67. Warren S. McCulloch, "What Is a Number, That a Man May Know It, and a Man, That He May Know a Number?" in Warren S. McCulloch, *Embodiments of Mind* (Cambridge MA: MIT Press, 1965), 1–18; 8.

68. For an historical and cultural analysis of the psychon, see Melissa M. Littlefield, "Matter for Thought: The Psychon in Neurology, Psychology, and American Culture, 1927–1943," in *Neurology and Modernity,* ed. Laura Salisbury and Andrew Shail (New York: Palgrave, 2010), 267–286.

69. Littlefield, "Matter for Thought."

70. Warren S. McCulloch, "What Is a Number," 8.

71. Warren S. McCulloch, "What's in the Brain That Ink May Character?" In Warren S. McCulloch, *Embodiments of Mind* (Cambridge, MA: MIT Press, 1965). (Originally presented at the International Congress for Logic, Methodology, and Philosophy of Science, Jerusalem, Israel, 28 August 1964, 387–397; 392–393.)

72. McCulloch, "What Is a Number."

73. Edgar D. Adrian, "On the Conduction of Subnormal Disturbances in Normal Nerve," *Journal of Physiology* 45 (1912): 389–412.

74. Edgar D. Adrian, "The All-or-None Principle in Nerve," *Journal of Physiology* 47 (1914): 460–474, 473.

75. Edgar D. Adrian and Yngve Zotterman, "The Impulses Produced by Sensory Nerve Endings, Part 2: The Response of a Single End-Organ," *Journal of Physiology* 61 (1926): 151–171. For more on inscription devices as key to the construction of scientific facts, see Bruno Latour and Steve Woolgar, *Laboratory Life: The Construction of Scientific Facts* (Beverly Hills: Sage, 1979), Chapter 2.

76. Frank, "Instruments," 233.

77. G. T. Kneebone, *Mathematical Logic and the Foundations of Mathematics* (London: D. van Nostrand, 1963), 5.

78. George Boole, *An Investigation of the Laws of Thought* (London, 1854).

79. E.g., Rafael Lorente de Nó, "Studies on the Structure of the Cerebral Cortex," *Journal für Psychologie und Neurologie* 45 (1933): 381–438; Rafael Lorente de Nó, "Facilitation of Motoneurones," *American Journal of Physiology* 113 (1935): 505–22. For more on Lorente de Nó, see Katharina Schmidt-Brücken, *Hirnzirkel: Kreisende Prozesse in Computer und Gehirn: Zur neurokybernetischen Vorgeschichte der Informatik* (Bielefeld: Transcript, 2012).

80. Samuel H. Bartley and George H. Bishop, "Factors Determining the Form of the Electrical Response from the Optic Cortex of the Rabbit," *American Journal of Physiology* 103 (1933): 173–184; Rafael Lorente de Nó, "Studies on the Structure of the Cerebral Cortex I. The Area Entorhinalis," *Journal für Psychologie und Neurologie* 45, no. 6 (1933): 381–438.

81. Boring, *History of Experimental Psychology* (New York: Century Co., 1929), 66.

82. Warren S. McCulloch to Lawrence S. Kubie, ca. May 1953, Folder "Kubie, Lawrence S.," Series I, WSM Papers, APS.

83. Lawrence S. Kubie, "A Theoretical Application to Some Neurological Problems of the Properties of Excitation Waves Which Move in Closed Circuits," *Brain* 53 (1930): 166–177.

84. McCulloch and Pitts, "A Logical Calculus," 115.

85. McCulloch and Pitts, "A Logical Calculus," 118.

86. McCulloch and Pitts, "A Logical Calculus," 117.

87. Gualtiero Piccinini, "The First Computational Theory of Brain and Mind: A Close Look at McCulloch and Pitts's 'A Logical Calculus of Ideas Immanent in Nervous Activity'," *Synthese* 141, no. 2 (2004): 175–215.

88. McCulloch and Pitts, "A Logical Calculus," 117.

89. McCulloch and Pitts, "A Logical Calculus," 132.

90. Christopher Lawrence and Anna-K. Mayer, "Regenerating England: An Introduction," in Christopher Lawrence and Anna-K. Mayer, *Regenerating England: Science, Medicine, and Culture in Interwar Britain* (Amsterdam, Atlanta, GA: Rodopi, 2000), 1–23; 1–3; Anna-K. Mayer, "'A Combative Sense of Duty': Englishness and the Scientists," in Lawrence and Mayer, *Regenerating England*, 67–106, 94.

91. Charles S. Sherrington, *Man on His Nature. The Gifford Lectures, Edinburgh, 1937–8* (Cambridge University Press, 1940), Chapters VII, VIII, IX, and XI.

92. Roger Smith, "Representations of Mind: C. S. Sherrington and Scientific Opinion, c. 1930–1950," *Science in Context* 14, no. 4 (2001): 511–539; Roger Smith, "Physiology and Psychology, or Brain and Mind, in the Age of C. S. Sherrington," in *Psychology in Britain: Historical Essays and Personal Reflections*, ed. Geoffrey C. Bunn, A. D. Lovie, and G. D. Richards (Leicester: British Psychological Society, 2001), 223–42.

93. Smith, "Representations of Mind."

94. Smith, "Representations of Mind," 525.

95. Smith, "Representations of Mind," 523.

96. Smith, "Representations of Mind," 533; Sherrington, *Man on his Nature*, 228.

97. Warren S. McCulloch, "Discussion by Dr. McCulloch of Paper by Dr. Alexander on Fundamental Concepts of Psychosomatic Research, Illinois Psychiatric Society, May 22, 1943," Series II, WSM Papers, APS.

98. McCulloch, "Discussion by Dr. McCulloch of Paper by Dr. Alexander."

99. Andrew Jewett, *Science, Democracy, and the American University: From the Civil War to the Cold War* (Cambridge University Press, 2012), 166.

100. Warren S. McCulloch to Francis J. Gerty, 10 November 1943, Folder "Gerty, Francis J./II," Series I, WSM Papers, APS.

101. Ralph S. Lillie to Warren S. McCulloch, 2 February 1943, Folder "Lillie, Ralph S.," Series I, WSM Papers, APS.

102. Lillie had recently published several papers in *Philosophy of Science* and *Journal of Philosophy* which he also sent to McCulloch. E.g., Ralph S. Lillie "Types of Physical Determination and the Activities of Organisms," *Journal of Philosophy* 28, no. 21 (1931): 561–573.

103. Warren S. McCulloch to Ralph S. Lillie, 10 February 1943, Folder "Lillie, Ralph S.," Series I, WSM Papers, APS.

104. Nadine Weidman, *Constructing a Scientific Psychology: Karl Lashley's Mind-Brain Debates* (Cambridge University Press, 1999), 1; Roger Smith, "Review of Kenton

Kroker, *The Sleep of Others and the Transformation of Sleep Research*," *History of the Human Sciences* 22, no. 5 (2009): 108–113, 108.

105. McCulloch and Pitts, "A Logical Calculus," 132.

106. Ibid.

## Chapter 5

1. E.g., Chester W. Darrow, Warren S. McCulloch, J. R. Greene, E. W. Davis, and Hugh W. Garol, "Parasympathetic Regulation of High Potential in the Electroencephalogram," *Journal of Neurophysiology* 7 (1944): 217–226; Warren S. McCulloch, H. B. Carlson, and Franz G. Alexander, "Zest and Carbohydrate Metabolism," *Research Publications of the Association for Research in Nervous and Mental Disease* XXIX (December 1949): 406–411; Ladislas J. Meduna and Warren S. McCulloch, "The Modern Concept of Schizophrenia," *The Medical Clinics of North America, Chicago Number* (W. B. Saunders, January 1945), 147–164; A. A. Ward, Warren S. McCulloch, and N. Kopeloff, "Temporal and Spatial Distribution of Changes during Spontaneous Seizures in Monkey Brain," *Journal of Neurophysiology* 11 (1948): 377–386; M. D. Wheatley and Warren S. McCulloch, "Sundry Changes in Physiology of Cerebral Cortex Following Rapid Injection of Sodium Cyanide," *Federal Proceedings of the American Societies for Experimental Biology* 6, no. 1 (1947): 225.

2. John Von Neumann, "First Draft Report on the EDVAC," report prepared for the U.S. Army Ordnance Department under contract W-670-ORD-4926, 1945, in N. Stern, *From ENIAC to UNIVAC* (Bedford, MA: Digital Press, 1981), 177–246. John von Neumann, *The Computer and the Brain (Yale University, Mrs. Hepsa Ely Silliman Memorial Lectures, 1958)* (New Haven and London: Yale University Press, 2000).

3. Hans Pols, "War Neurosis, Adjustment Problems in Veterans, and an Ill Nation: The Disciplinary Project of American Psychiatry during and after World War II," *Osiris* Second Series 22 (2007): 72–92.

4. Gerald N. Grob, "The Forging of Mental Health Policy in America: World War II to the New Frontier," *Journal of the History of Medicine and Allied Sciences* 42 (1987): 410–446; 413.

5. Pols, "War Neurosis."

6. Elizabeth Lunbeck, *The Psychiatric Persuasion: Knowledge, Gender, and Power in Modern America* (Princeton, NJ: Princeton University Press, 1994), 306–307.

7. Hans Pols, "War Neurosis."

8. Gerald N. Grob, *Mental Illness and American Society, 1875–1940* (Princeton, NJ: Princeton University Press, 1983), Chapter 11. For histories of electroconvulsive therapy, see e.g., Laura Hirshbein, "Historical Essay: Electroconvulsive Therapy,

Memory, and Self in America," *Journal of the History of the Neurosciences* 21 (2012): 147–169; Laura Hirshbein and Sharmalie Sarvananda, "History, Power, and Electricity: American Popular Magazine Accounts of Electroconvulsive Therapy, 1940–2005," *Journal of the History of the Behavioral Sciences* 44, no. 1 (Winter 2008): 1–18; Edward Shorter and David Healy, *Shock Therapy: A History of Electroconvulsive Treatment in Mental Illness* (Toronto: University of Toronto Press, 2007); Timothy W. Kneeland and Carol A. B. Warren, *Pushbutton Psychiatry: A History of Electroshock in America* (Westport, CT: Praeger, 2002).

9. Joel T. Braslow, *Mental Ills and Bodily Cures: Psychiatric Treatment in the First Half of the Twentieth Century* (Berkeley: University of California Press, 1997); Sarah Linsley Starks and Joel T. Braslow, "The Making of Contemporary American Psychiatry, Part 1: Patients, Treatments, and Therapeutic Rationales before and after World War II," *History of Psychology* 8, no. 2 (2005): 176–193.

10. For the context of schizophrenia during the early twentieth century, see Richard Noll, *American Madness: The Rise and Fall of Dementia Praecox* (Cambridge, MA: Harvard University Press, 2011), Chapter 8.

11. Grob, *Mental Illness and American Society*, 299.

12. Grob, *Mental Illness and American Society*, 297–305. Statisticians Benjamin Malzberg and Horatio M. Pollack conducted these studies: e.g., Benjamin Malzberg, "Outcome of Insulin Treatment of One Thousand Patients with Dementia Praecox," *Psychiatric Quarterly* 12 (1938): 528–53; Horatio M. Pollock, "A Statistical Study of 1,140 Dementia Praecox Patients Treated with Metrazol," *Psychiatric Quarterly* 13 (1939): 558–68.

13. Frankwood E. Williams to Dr. Frank Norbury, 14 November 1935, quoted in Gerald N. Grob, *The Inner World of American Psychiatry, 1890–1940: Selected Correspondence* (New Brunswick, NJ: Rutgers University Press, 1985), 220–222.

14. Gerald N. Grob, "World War II and American Psychiatry," *Psychohistory Review* 19 (1990): 41–69.

15. Nathan G. Hale, Jr., *The Rise and Crisis of Psychoanalysis in the United States: Freud and the Americans, 1917–1985* (New York, Oxford: Oxford University Press, 1995), 200.

16. Gerty to Alan Gregg, 11 October 1944, RG 1.1 Series 200A, Box 112.1374, RAC. In April 1951, Robert Morison, who had been Assistant Director of the Medical Sciences Division since 1944, continued Gregg's tradition. R. S. Morison to A.J. Warren, 11 April 1951, RG 3, Series 906, Box 1.5, RAC.

17. Report, April 1 1950. Rockefeller Foundation Trustees Bulletin 1950, RAC.

18. Morison Interview, 25 February 1947.

19. Pitts to McCulloch, 26 October 1943, Folder "Pitts, Walter," Series I, WSM Papers, APS.

20. Mary Jean Vasiloff, Personal Communication, 5 August 2015.

21. The Josiah Macy Jr. Foundation, *The Josiah Macy Jr. Foundation 1930–1955. A Review of Activities* (New York: The Josiah Macy Jr. Foundation, 1955), 5.

22. Josiah Macy Jr. Foundation. *Twentieth Anniversary Review of the Josiah Macy Jr. Foundation* (New York: The Josiah Macy Jr. Foundation, 1950), 17 (from a letter of gift written by Kate Macy Ladd on April 24, 1930, when she established the foundation as a memorial to her father, Josiah Macy, Jr.).

23. Warren S. McCulloch, "A query as to the possibility of obtaining assistants in technical procedures in the Research Laboratories of the Illinois Neuropsychiatric Institute," 16 November 1942, Folder "Gerty, Francis J./III," Series I, WSM Papers, APS.

24. E.g., Gerhardt von Bonin, Hugh W. Garol, and Warren S. McCulloch, "The Functional Organization of the Occipital Lobe," *Biological Symposia* VII (1942): 164–192; Percival Bailey, Gerhardt von Bonin, E. W. Davis, Hugh W. Garol, Warren S. McCulloch, E. Roseman, and A. Silveira, "Functional Organization of the Medial Aspect of the Primate Cortex," *Journal of Neurophysiology* 7 (1944): 51–56; Warren S. McCulloch, C. Graf, and Horace W. Magoun, "A Cortico-bulbo-reticular Pathway from Area 4-s," *Journal of Neurophysiology* 9 (1946): 127–132.

25. Warren S. McCulloch, "The Functional Organization of the Cerebral Cortex," *Physiological Reviews* 24, no. 3 (1944): 390–407.

26. Chester W. Darrow, J. R. Green, and Warren S. McCulloch, "Activity in the Great Superficial Petrosal Nerve Influencing the Electroencephalogram," *Federation Proceedings* 3, no. 1 (1944): 8.

27. Arthur A. Ward, Warren S. McCulloch, and Horace W. Magoun, "Production of an Alternating Tremor at Rest in Monkeys," *Journal of Neurophysiology* 11 (1948): 317–330.

28. Fremont-Smith to McCulloch, 4 January 1944, Folder "Fremont-Smith/I," Series I, WSM Papers, APS. McCulloch also participated in a June 1945 similarly organized conference (suggested by Office of Air Surgeon, Army Air Forces) on "The Teaching of Psychotherapy." Fremont-Smith to McCulloch, 15 May 1945, Folder "Fremont-Smith/I," Series I, WSM Papers, APS.

29. Francis J. Gerty and McCulloch to Fremont-Smith, 17 February 1944, Folder "Frank Fremont-Smith/I," Series I, WSM Papers, APS, Macy Foundation Zest Metabolism Fund, 1 July 1945, Folder "Josiah Macy Jr. Foundation #1," Series II, WSM Papers, APS.

30. Frank Fremont-Smith to McCulloch, 4 February 1944, Folder "Fremont-Smith, Frank," Series I, WSM Papers, APS.

31. McCulloch to Fremont-Smith, 13 June 1944, Folder "Fremont-Smith/I," Series I, WSM Papers, APS.

32. Warren S. McCulloch to Francis J. Gerty, 25 August 1943, Folder "Gerty, Francis J./II," Series I, WSM Papers, APS.

33. Ladislas von Meduna, *Die Konvulsiontherapie der Schizophrenia* (Halle: Carl Marhold, 1937).

34. Grob, *Mental Illness and American Society*, 299.

35. Frank Fremont-Smith to McCulloch, 22 September 1944, Folder "Fremont-Smith, Frank," Series I, WSM Papers, APS. The project was "Process of Schizophrenia and Allied Conditions," by Gerty, Meduna, and McCulloch. This was a $7,000 grant for a year from 1944 to 1945. Macy Foundation Schiz Metabolism Fund, 1 July 1945, Folder "Josiah Macy Jr. Foundation #1," Series II, WSM Papers, APS.

36. Meduna and McCulloch, "The Modern Concept of Schizophrenia."

37. "The Illinois Neuropsychiatric Institute," excerpt from Trustees Confidential Bulletin, February 1949, RG 1.1 Series 200A, Box 112.1376, RAC.

38. McCulloch to Fremont-Smith, 1 June 1948, RG 1.1 Series 200A, Box 112.1375, RAC.

39. L. J. Meduna to Alan Gregg, 21 January 1949, RG 1.1 Series 200A, Box 112.1376, RAC.

40. Gerty to Gregg, 11 October 1944. RG 1.1 Series 200A, Box 112.1374, RAC.

41. E. W. Davis, Warren S. McCulloch, and E. Roseman, "Rapid Changes in the $O_2$ Tension of Cerebral Cortex during Induced Convulsions," *American Journal of Psychiatry* 100 (1944): 825–829.

42. McCulloch to Fremont-Smith, 1 June 1948. M. D. Wheatley and Warren S. McCulloch, "Sundry Changes in Physiology of Cerebral Cortex."

43. "The Illinois Neuropsychiatric Institute," excerpt from Trustees Confidential Bulletin, February 1949, RG 1.1 Series 200A, Box 112.1376, RAC.

44. Max Fink, "Images in Psychiatry: Ladislas J. Meduna, M.D. 1896–1964," *American Journal of Psychiatry* 156, no. 11 (November 1999): 1807.

45. Quarterly Report, 1 July 1949 through 30 September 1949, Navy Contract N6ori-07125, Folder "United States. Naval Medical Research Laboratory," Series I, WSM Papers, APS.

46. Robert S. Morison Interview, 19 September 1946, RG 1.1, Series 200A, Box 112.1374, RAC.

47. Andrew Scull, *Madhouse: A Tragic Tale of Melagomania and Modern Medicine* (New Haven and London: Yale University Press, 2005), 283.

48. E.g., Elliot Valenstein, *Great and Desperate Cures: The Rise and Decline of Psychosurgery and Other Radical Treatments for Mental Illness* (New York: Basic Books, 1986).

49. Jack Pressman has eloquently and masterfully achieved this in his history of lobotomy: *Last Resort: Psychosurgery and the Limits of Medicine* (Cambridge: University of Cambridge Press, 1998). See also Braslow, *Mental Ills and Bodily Cures*.

50. Braslow, *Mental Ills and Bodily Cures*, 9.

51. "New Horizons of Psychiatry," *New York Times*, 3 January 1948, 12.

52. Robert S. Morison Interview, 25 February 1947, RG 1.1, Series 200A, Box 112.1375, RAC.

53. Morison Interview, 25 February 1947.

54. Undated document, Folder Academic Laboratory Staff, WSM Papers, No. 2.

55. McCulloch to Fremont-Smith, 13 June 1944, Folder "Fremont-Smith/I," Series I, WSM Papers, APS.

56. Robert S. Morison Interview, 25 February 1947, RG 1.1, Series 200A, Box 112.1375, RAC.

57. Handwritten discussion notes, n.d., Folder "American Society for Research in Psychosomatic Problems," Series II, WSM Papers, APS.

58. Warren S. McCulloch, "Speech for the Association for Psychoanalytic and Psychosomatic Medicine," n.d. (c. 1945), Folder "Association for Research in Psychosomatic Problems," Series II, WSM Papers, APS.

59. McCulloch, "Speech for the Association for Psychoanalytic and Psychosomatic Medicine."

60. McCulloch, "Speech for the Association."

61. McCulloch, "Speech for the Association."

62. McCulloch to Fremont-Smith, 27 August 1945, Folder "Fremont-Smith, Frank," Series I, WSM Papers, APS.

63. Warren S. McCulloch, "Modes of Functional Organization in the Cerebral Cortex." *Federation Proceedings*, 6, no. 2 (June 1945): 448–552; reprinted in *The Collected Works of Warren S. McCulloch, Volume 3*, ed. Rook McCulloch (Salinas, CA: Intersystems, 1989), 500–510.

64. McCulloch, "Modes of Functional Organization in the Cerebral Cortex," 501.

65. John Hughlings Jackson, Croonian Lecture, Part 1, March 1884 (*Lancet* 1 [1884]: 535).

66. McCulloch, "Modes of Functional Organization," 506.

67. Warren S. McCulloch, "A Recapitulation of the Theory, with Forecast of Several extensions," *Annals of the New York Academy of Sciences* 50, no. 4 (13 October 1948): 259–288.

68. McCulloch, "A Recapitulation of the Theory," 565.

69. Warren S. McCulloch, "Physiological Processes Underlying Psychoneuroses," *Proceedings of the Royal Society, Medical Supplement* XLII (Anglo-American Symposium on Psychosurgery, Neurophysiology, and Physical Treatments in Psychiatry) (1949): 72–84.

70. Warren S. McCulloch, "The Brain as a Computing Machine," in *The Collected Works of Warren S. McCulloch, Volume 2*, ed. Rook McCulloch (Salinas, CA: Intersystems, 1989), 582–596.

71. Rebecca Jo Plant, "William Menninger and American Psychoanalysis, 1946–48," *History of Psychiatry* 16, no. 2 (2005): 181–202. For more on this expansion, see Hale, *The Rise and Crisis of Psychoanalysis*; John C. Burnham (ed.), *After Freud Left: A Century of Psychoanalysis in America* (Chicago and London: University of Chicago Press, 2012); and Eli Zaretsky, *Secrets of the Soul: A Social and Cultural History of Psychoanalysis* (New York: Knopf, 2004).

72. Hale, *The Rise and Crisis of Psychoanalysis*, 208.

73. Plant, "William Menninger," 197.

74. Warren S. McCulloch, "The Past of a Delusion," Paper Presented to the Chicago Literary Club, January 1952, in *The Collected Works of Warren S. McCulloch, Volume 2* (Salinas, CA: Intersystems, 1989), 761–791; 786. For McCulloch's disagreements with Kubie on psychoanalysis see Heims, *The Cybernetics Group*, 120–146.

75. McCulloch, "The Past of a Delusion," 764.

76. Jonathan Sadowsky, "Beyond the Metaphor of the Pendulum: Electroconvulsive Therapy, Psychoanalysis, and the Styles of American Psychiatry," *Journal of the History of Medicine and Allied Sciences* 61, no. 1 (2005): 1–25.

77. Sadowsky, "Beyond the Metaphor of the Pendulum," 13; Frank Sulloway, *Freud: Biologist of the Mind* (New York: Basic Books, 1979).

78. E.g., Andrew Pickering, "Cyborg History and the World War II Regime," *Perspectives on Science* 3 (1995): 1–48; Paul N. Edwards, *The Closed World: Computers and the Politics of Discourse in Cold War America* (Cambridge, MA: MIT Press, 1996);

N. Katherine Hayles, *How We Became Posthuman: Virtual Bodies in Cybernetics, Litera-
ture, and Informatics* (Chicago: University of Chicago Press, 1999); Lily E. Kay, *Who
Wrote the Book of Life? A History of the Genetic Code* (Stanford: Stanford University
Press, 2000).

79. "The Illinois Neuropsychiatric Institute," excerpt from Trustees Confidential
Bulletin, February 1949, RG 1.1, Series 200A, Box 112.1376, RAC.

80. E.g., Warren S. McCulloch and Walter Pitts, "How Nervous Structures Have
Ideas"; Warren S. McCulloch, "Machines That Think and Want."

81. Morison to Frank E. Mott, 28 July 1947, RG 1.1, Series 200A Box 112.1375, RAC.

## Chapter 6

1. Memo, "A Meeting for Rapproachment of Mathematical Methods and Electrophysiology of the Central Nervous System," Folder "Josiah Macy Jr. Foundation,"
Series II, WSM Papers, APS.

2. It is likely that this was the plan for the First Macy Conference on Cybernetics,
held in March 1946. According to Steve J. Heims, the first Macy meeting may
have simply included laboratory neurophysiology, psychiatry, and mathematics
had it not been for Gregory Bateson's urging that social and behavioral scientists
be included (Steve J. Heims, *The Cybernetics Group* [Cambridge, MA: MIT Press,
1993], 17).

3. Wiener to Rosenblueth, 24 January 1945, Box 4, Folder 67, NW Papers, MITASC.

4. Wiener to Rosenblueth, 24 January 1945.

5. Peter Galison "The Ontology of the Enemy: Norbert Wiener and the Cybernetic
Vision," *Critical Inquiry* 21, no. 1 (1994): 228–266; Andrew Pickering, "Cyborg His-
tory and the World War II Regime," *Perspectives on Science* 3 (1995): 1–48; Heims,
*The Cybernetics Group*; N. Katherine Hayles, *How We Became Posthuman: Virtual
Bodies in Cybernetics, Literature, and Informatics* (Chicago: University of Chicago
Press, 1998); Paul N. Edwards, *The Closed World: Computers and the Politics of Discourse in Cold War America* (Cambridge, MA: MIT Press, 1996); Geof Bowker, "How
to Be Universal: Some Cybernetic Strategies, 1943–1970," *Social Studies of Science* 23
(1993): 107–127.

6. Wiener to Rosenblueth, 11 August 1945, 20 August 1945, Box 4, Folder 68, NW
Papers, MITASC.

7. Bowker, "How to Be Universal"; Peter Galison, "The Americanization of Unity,"
in *Science in Culture*, ed. Peter Galison, S.R. Graubard, and Everett Mendelsohn (New
Brunswick & London: Transaction Publishers, 2001), 45–71.

8. Pnina Abir-Am, "The Biotheoretical Gathering, Transdisciplinary Authority, and the Incipient Legitimation of Molecular Biology in the 1930s: New Perspective on the Historical Sociology of Science," *History of Science* 25, 67 (1987): 1–70; Lily E. Kay, *The Molecular Vision of Life: Caltech, The Rockefeller Foundation, and the Rise of the New Biology* (Oxford University Press, 1993); Lily E. Kay. *Who Wrote the Book of Life? A History of the Genetic Code* (Stanford: Stanford University Press, 2000); Evelyn Fox Keller, *Refiguring Life: Metaphors of Twentieth-Century Biology* (New York: Columbia University Press, 1995).

9. Robert E. Kohler, *From Medical Chemistry to Biochemistry: The Making of a Biomedical Discipline* (Cambridge: Cambridge University Press, 1982), 1.

10. Adele Clarke and Joan Fujimura, "Which Tools? Which Jobs? Why Right?" 16.

11. Ronald Kline, "Where Are the Cyborgs in Cybernetics?" *Social Studies of Science*, 33 (June 2009): 331–362.

12. William Eckhardt, "Changing Concerns in Peace Research and Education," *Bulletin of Peace Proposals* 3 (1974): 280.

13. Julie Thompson Klein, "The Transdisciplinary Moment(um)," *Integral Review* 9, no. 2 (June 2013): 189–199.

14. Leah Ceccarelli, "Polysemy: Multiple Meanings in Rhetorical Criticism," *Quarterly Journal of Speech* 84, no. 4 (November 1998): 395–415; Leah Ceccarelli, *Shaping Science with Rhetoric: The Cases of Dobzhansky, Schrödinger, and Wilson* (Chicago: University of Chicago Press, 2001), 162–165, 180–81.

15. Andrew Pickering, *The Cybernetic Brain: Sketches of Another Future* (Chicago: University of Chicago Press, 2010); Andrew Pickering, "Ontology and Antidisciplinarity," in *Interdisciplinarity: Reconfigurations of the Social and Natural Sciences*, ed. Andrew Barry and Georgina Born, 209–225 (London: Routledge, 2013).

16. Galison, "The Americanization of Unity."

17. Richardson's account of the history of scientific philosophy emphasizes the point that logical empiricism had much more to do with reforming philosophy itself rather than improve science (Alan Richardson, "Scientific Philosophy as a Topic for History of Science," *Isis* 99 [2008]: 88–96).

18. For more on the context of these meetings in Harvard's "interstitial academy," see Joel Isaac, *Working Knowledge: Making the Human Sciences from Parsons to Kuhn* (Cambridge, MA: Harvard University Press, 2012), 149–151.

19. Boring to Wiener, 8 February 1945, Box 4, Folder 67, NW Papers, MITASC; Gerald Holton, "On the Vienna Circle in Exile: An Eyewitness Report," in *The Foundational Debate: Complexity and Constructivity in Mathematics and Physics*, ed. Werner Depauli-Schimanovich, Eckehart Köhler, and Friedrich Stadler (Dordrecht: Kluwer, 1995), 269–292.

20. Gary L. Hardcastle, "Debabelizing Science: The Harvard Science of Science Discussion Group, 1940-44," in *Logical Empiricism in North America*, ed. Gary L. Hardcastle and Alan Richardson (Minneapolis, MN: University of Minnesota Press, 2003), 170–196.

21. Gerald Holton, "On the Vienna Circle in Exile," 276.

22. Holton, "On the Vienna Circle in Exile," 275. See also Isaac, *Working Knowledge*, Chapter 5.

23. Galison, "The Americanization of Unity."

24. Philipp Frank, "The Institute for the Unity of Science: Its Background and Purpose," *Synthese* 6 (3/4 – July/August 1947): 160–167, 160.

25. George A. Reisch, *How the Cold War Transformed Philosophy of Science: To the Icy Slopes of Logic* (Cambridge: Cambridge University Press, 2005), viii, 54–55. However, as Reisch notes, ultimately the Unity of Science movement, and philosophy of science more broadly, became "depoliticized" during the Cold War due to political pressures in the anticommunist climate of McCarthyism. The "collectivist" spirit of the unity of science promoters was seen to be out of step with the political climate that defined Cold War America (Reisch, *How the Cold War Transformed the Philosophy of Science*, 19–21).

26. Reisch, *How the Cold War Transformed the Philosophy of Science*, 296; see Philipp Frank, "Introductory Remarks," *Contributions to the Analysis and Synthesis of Knowledge, Proceedings of the American Academy of Arts and Sciences* 80 (1951): 5–8. Part of this drive was an interest in the new "science of man"; see Reisch, *How the Cold War Transformed Philosophy of Science*, 297.

27. Galison, "The Americanization of Unity," 47–48; Hardcastle, "Debabelizing Science," 186–188.

28. Frank, "Communications Engineering and Theory" (1952) RG 1.1, Series 100 Unity of Science, 1952–56, Box 35.285, RAC.

29. Norbert Wiener, *Cybernetics: Or Control and Communication in the Animal and the Machine* (Cambridge, MA: MIT Press, 1948), Introduction; Arturo Rosenblueth and Norbert Wiener, "The Role of Models in Science," *Philosophy of Science* 12 (1945): 316–321.

30. E.g., Heims, *The Cybernetics Group*; Bowker, "How to Be Universal"; Jean-Pierre Dupuy, *The Mechanization of the Mind: On the Origins of Cognitive Science*, trans. M.B. DeBevoise (Princeton, NJ: Princeton University Press, 2000); Galison, "The Ontology of the Enemy"; David Mindell, Jérôme Ségal, and Slava Gerovitch, "From Communications Engineering to Communications Science: Cybernetics and Information Theory in the United States, France, and the Soviet Union," in *Science and Ideology: A Comparative History*, ed. Mark Walker (London: Routledge, 2003), 66–95; Wiener,

Cybernetics. Although Wiener built up this observation in his own cybernetics origin stories as monumental, historian of technology David Mindell has illuminated the multiple traditions of "feedback" in the context of technological and engineering practice during World War I and through to the World War II. See David Mindell, *Between Human and Machine: Feedback, Control, and Computing Before Cybernetics* (Baltimore: Johns Hopkins University Press, 2002).

31. See, for example, Norbert Wiener and Arturo Rosenblueth, "The Mathematical Formulation of the Problem of Conduction of Impulses in a Network of Connected Excitable Elements, Specifically in Cardiac Muscle," *Archives of the Institute for Cardiology, Mexico* (1946): 205–265. I discuss their collaboration in more detail in Tara H. Abraham, "Transcending Disciplines: Scientific Styles in Studies of the Brain in Mid-Twentieth Century America," *Studies in History and Philosophy of Biological and Biomedical Sciences* 43 (2012): 552–568.

32. Rosenblueth to Wiener, 26 June 1945, Box 4, Folder 68, NW Papers, MITASC.

33. Rosenblueth and Wiener, "The Role of Models in Science," 316.

34. Wiener to Hoagland, 24 January 1947, Box 5, Folder 74, NW Papers, MITASC.

35. Wiener, *Cybernetics*, 3.

36. Wiener, *Cybernetics*, 1.

37. Wiener, *Cybernetics*, 2.

38. Carl G. Hempel and Paul Oppenheim, "Studies in the Logic of Explanation," *Philosophy of Science* 15, no.1 (1948): 135–175, 144; Ernst Nagel, "Teleological Explanation and Teleological Systems," in *Readings in the Philosophy of Science*, ed. Herbert Feigl and M. Brodbeck (New York: Appleton-Century-Crofts, 1953), 537–558; Richard B. Braithwaite, "Causal and Teleological Explanation," in *Scientific Explanation: A Study of the Function of Theory, Probability, and Law in Science* (Based upon the Tanner Lectures, 1946), (Cambridge University Press, 1964), 319–341; Paul Oppenheim and Hilary Putnam, "Unity of Science as a Working Hypothesis," in *Concepts, Theories, and the Mind–Body Problem, Minnesota Studies in the Philosophy of Science, Volume* 2, ed. Herbert Feigl, Michael Scriven, and Grover Maxwell (Minneapolis, MN: University of Minnesota Press, 1958), 3–36; 19–20.

39. *Twentieth Anniversary Review of the Josiah Macy Jr. Foundation* (New York: Josiah Macy Jr. Foundation, 1950), 19. For more on the increase in federal funds for medicine, see Stella Leche Deignan and Esther Miller, "The Support of Research in Medical and Allied Fields for the Period 1946 through 1951," *Science* 115, no. 2987 (March 28, 1952): 321–343.

40. *The Josiah Macy Jr. Foundation: A Review of Activities* (New York: Josiah Macy Jr. Foundation, 1955), 8.

41. Ibid.

42. *Twentieth Anniversary Review of the Josiah Macy Jr. Foundation* (New York: Josiah Macy Jr. Foundation, 1950), 87.

43. *Twentieth Anniversary Review,* 88.

44. Heinz Von Foerster, "Circular Causality: The Beginnings of an Epistemology of Responsibility," in *The Collected Works of Warren S. McCulloch, Volume 3,* ed. Rook McCulloch (Salinas, CA: Intersystems, 1989), 808–829, 810.

45. Frank Fremont-Smith. "Introductory Discussion," in *Cybernetics: Circular Causal, and Feedback Mechanisms in Biological and Social Systems: Transactions of the Sixth Conference,* ed. Heinz von Foerster (New York: Josiah Macy Jr. Foundation, 1950), 9–26, 9. In Claus Pias (ed.) *Cybernetics—Kybernetik: The Macy Conferences 1946–1953* (Zürich-Berlin: Diaphanes, 2003), 29–40, 29.

46. Ibid.

47. Ibid.

48. Fremont-Smith to McCulloch, 8 February 1946, Folder "Conference: Josiah Macy Meeting I," Series II, WSM Papers, APS.

49. Indeed, Fremont-Smith's welcoming remarks to a small conference in September 1946 said that "Dr. McCulloch was instrumental" in encouraging the Macy Foundation to hold the conference on circular mechanisms [held in March 1946]. "Conference on Teleological Mechanisms September 20, 1946" Transcript, Folder "Conference: Cybernetics," Series II, WSM Papers, APS.

50. "Conference on Feedback Mechanisms and Circular Causal Systems in Biology and the Social Sciences," n.d., Folder "Conference: Josiah Macy Meeting I," Series II, WSM Papers, APS.

51. McCulloch to von Neumann, 22 February 1946, Folder "Conference: Josiah Macy Meeting I," Series II, WSM Papers, APS.

52. McCulloch to Bateson and Mead, 11 February 1946, Folder "Conference: Josiah Macy Meeting I," Series II, WSM Papers, APS.

53. Warren S. McCulloch to Kurt Lewin, 15 November 1946, Folder "Conference: Josiah Macy Meeting III," Series II, WSM Papers, APS.

54. Lawrence K. Frank, "Foreword," *Annals of the New York Academy of Sciences,* Volume 50: Teleological Mechanisms (1948), 189–196, 189.

55. Ibid., 191.

56. Heims, *The Cybernetics Group,* 37.

57. David McCulloch, Personal Communication, 4 February 2015.

58. Lawrence K. Frank to McCulloch, 7 October 1946, Folder "Conference: Josiah Macy Meeting II," Series II, WSM Papers, APS.

59. McCulloch's Summary of the First Three Meetings, quoted in Heims, *The Cybernetics Group*, 205.

60. Rosenblueth to McCulloch, 3 February 1947, Folder "Conference: Josiah Macy Foundation Meeting IV," Series II, WSM Papers, APS.

61. Morison Diary, 12 March 1946, Record Group 12.1 Box 46 Folder Jan–June 1946, RAC.

62. Robert S. Morison Interview, 2 May 1946, RG 1.1, Series 200A, Box 112.1374, RAC.

63. Morison to Rosenblueth, 17 January 1947, Record Group 1.1 Series 224, Box 1.3, RAC.

64. Rosenblueth to Morison, 25 January 1947, Record Group 1.1 Series 224, Box 1.3, RAC.

65. See letters Wiener to Pitts 4 and 5 April 1947, McCulloch to Wiener 8 April 1947, Rosenblueth to Wiener 9 April 1947, Rosenblueth to Wiener 10 April 1947, Wiener to Pitts 10 April 1947, Wiener to Rosenblueth 10 April 1947, Pitts to Wiener 11 April 1947, McCulloch to Lettvin 15 April 1947, Wiener to Rosenblueth 16 April 1947, Wiener to Pitts 16 April 1947. Box 5 Folder 77 NW Papers, MITASC.

66. Flo Conway and Jim Siegelman, *Dark Hero of the Information Age: In Search of Norbert Wiener, Father of Cybernetics* (New York, Basic Books, 2005), 225–227.

67. Molly Harrower to McCulloch, 14 January 1948, Folder "Conference: Josiah Macy Foundation Meeting V," Series II, WSM Papers, APS.

68. Margaret Mead, "Possible Mechanisms of Recall and Recognition," in *Cybernetics: Circular Causal, and Feedback Mechanisms in Biological and Social Systems*, ed. Heinz von Foerster (New York: Josiah Macy Jr. Foundation, 1950), 146–202, 148. In Claus Pias (ed.) *Cybernetics—Kybernetik: The Macy Conferences 1946–1953* (Zürich-Berlin: Diaphanes, 2003), 122–162, 123.

69. McCulloch to Fremont-Smith, 8 January 1949, Folder "Conference: Josiah Macy Foundation Meeting VI, #2," Series II, McCulloch Papers, APSL.

70. Von Foerster, "Circular Causality."

71. Transcript of Chairman and Editors Meeting, 27 April 1949, Folder: "Conference: Josiah Macy Foundation Meeting VI, #1," Series II, WSM Papers, APS.

72. Memo, "To the Members of the Feedback Conference on Teleological Mechanisms," n.d., Folder "Conference: Josiah Macy Foundation Meeting VI, #2," Series II, WSM Papers, APS.

73. John von Neumann to Norbert Wiener, 29 November 1946, Box 7, Folder "Wiener, Norbert," John Von Neumann Papers, Library of Congress, Washington DC.

74. Ralph W. Gerard, "Some of the Problems Concerning Digital Notions in the Central Nervous System," in *Cybernetics: Circular Causal, and Feedback Mechanisms in Biological and Social Systems: Transactions of the Seventh Conference*, ed. Heinz von Foerster, Margaret Mead, and Hans Lukas Teuber (New York: Josiah Macy Jr. Foundation, 1951), 11–57, 11. In Claus Pias (ed.) *Cybernetics—Kybernetik: The Macy Conferences 1946–1953* (Zürich-Berlin: Diaphanes, 2003), 171–202, 171.

75. See, for example, Howard Gardner, *The Mind's New Science: A History of the Cognitive Revolution* (New York: Basic Books, 1985), 10, 23; Barbara von Eckardt, *What Is Cognitive Science?* (Cambridge, MA: MIT Press, 1993); Bernard J. Baars, *The Cognitive Revolution in Psychology* (New York: Guilford Press, 1986).

76. Walter Pitts and Warren S. McCulloch, "How We Know Universals: The Perception of Auditory and Visual Forms," *Bulletin of Mathematical Biophysics* 9, no. 3 (1947): 127–147.

77. Kenneth J. W. Craik, *The Nature of Explanation* (Cambridge: Cambridge University Press, 1943); Alan F. Collins, "An Asymmetric Relationship: The Spirit of Kenneth Craik and the Work of Warren McCulloch," *Interdisciplinary Science Reviews* 37, no. 3 (2012): 254–268. For an analysis of the Pitts–McCulloch work from the perspective of visual representation, see Tara H. Abraham, "From Theory to Data: Representing Neurons in the 1940s," *Biology and Philosophy* 18, no. 3 (2003): 415–426.

78. Warren S. McCulloch, "Why the Mind Is in the Head," in *Cerebral Mechanisms in Behavior: The Hixon Symposium*, ed. Lloyd A. Jeffress (New York: John Wiley, 1951), 42–111; 51.

79. Pitts and McCulloch, "How We Know Universals," 146.

80. McCulloch, "Why the Mind Is in the Head," 53.

81. For the use of material models in cybernetic practice, see Peter Asaro, "Working Models and the Synthetic Method: Electronic Brains as Mediators between Neurons and Behavior," *Science Studies* 19, no. 1 (2006): 12–34.

82. McCulloch, "Why the Mind Is in the Head," 43.

83. Ibid., 46, 47.

84. For an overview of this discussion see Margaret Morrison and Mary S. Morgan, "Introduction," in *Models as Mediators: Perspectives on Natural and Social Science*, ed. Mary S. Morgan and Margaret Morrison (Cambridge: Cambridge University Press, 1999), 1–9.

85. E.g., Mary S. Morgan and Till Grüne-Yanoff, "Modelling Practices in the Social and Human Sciences: An Interdisciplinary Exchange," *Perspectives on Science* 21, no. 2 (2011): 143–156.

86. Margaret Morrison and Mary S. Morgan, "Models as Mediating Instruments" in *Models as Mediators: Perspectives on Natural and Social Science*, ed. Mary S. Morgan and Margaret Morrison (Cambridge: Cambridge University Press, 1999), 10–37.

87. Alastair C. Crombie, *Styles of Scientific Thinking in the European Tradition: The History of Argument and Explanation Especially in the Mathematical and Biomedical Sciences and Arts* (London: Duckworth, 1994). For more on the cybernetic style, see Abraham, "Transcending Disciplines." For more on the notion of scientific styles, see Ian Hacking, "'Style' for Historians and Philosophers," *Studies in the History and Philosophy of Science* 23 (1992): 1–20; Martin Kusch, "Hacking's Historical Epistemology: A Critique of Styles of Reasoning," *Studies in History and Philosophy of Science* 41 (2010): 158–173.

88. Mary S. Morgan, *The World in a Model: How Economists Work and Think* (Cambridge University Press, 2012), 20.

89. Morgan, *The World in a Model*, 21–22.

90. McCulloch, "Why the Mind Is in the Head," 42.

91. Kurt Danziger, *Naming the Mind: How Psychology Found Its Language* (London: Sage, 1997), 7.

92. *Transactions of the American Neurological Association 1952* (Seventy-sixth Annual Meeting, June 1951).

93. E.g., Edgar D. Adrian, Frederic Bremer, and Herbert H. Jasper (eds.), *Brain Mechanisms and Consciousness* (Oxford: Blackwell, 1954); P. Laslett, *The Physical Basis of Mind: A Symposium* (New York: Macmillan, Oxford: Basil Blackwell, 1950); C. Solomon, S. Cobb, and W. Penfield (eds.), *The Brain and Human Behavior: Procceedings of the Meeting of the Association for Research in Nervous and Mental Disease, 7 and 8 December, 1956* (Baltimore: Williams and Wilkins, 1958).

94. Horace W. Magoun, "An Ascending Reticular Activating System in the Brain Stem," *Archives of Neurology and Psychiatry* 67, no. 2 (1952): 145–154.

95. These had been anatomically demonstrated by Santiago Ramón y Cajal and physiologically demonstrated by Forbes, Cobb, Catell, and Lorente de Nó. In Cobb's view, Lorente de No's work on reverberating circuits provided the "data needed by Wiener, McCulloch, and Pitts to develop cybernetics" (Stanley Cobb, "On the Nature and Locus of Mind," *Archives of Neurology and Psychiatry* 67, no. 2 [1952]: 172–177, 173–174).

96. Pitts and McCulloch, "How We Know Universals."

97. Cobb, "On the Nature and Locus of Mind."

98. Ibid., 176.

99. E.g., S. Zuckerman in Laslett, *The Physical Basis of Mind*, 26; Edgar D. Adrian, *The Physical Background of Perception* (Oxford: Clarendon Press, 1947).

100. McCulloch and Pitts, "A Logical Calculus," 115.

101. McCulloch, "Why the Mind Is in the Head," 53.

102. Lashley in McCulloch, "Why the Mind Is in the Head," 70.

103. Lashley in McCulloch, "Why the Mind Is in the Head," 70–71.

104. Nadine M. Weidman, *Constructing Scientific Psychology: Karl Lashley's Mind-Brain Debates* (Cambridge: Cambridge University Press, 1999), 48–49.

105. Weidman, *Constructing a Scientific Psychology*, Chapter 7.

106. Karl Lashley, address to the American Neurological Association, June 1951, quoted by Stanley Cobb, "A Salute from Neurologists," in *The Neuropsychology of Lashley*, ed. Frank A. Beach, Donald O. Hebb, Clifford T. Morgan, and Henry W. Nissen (New York: McGraw Hill, 1960), xix.

107. Von Neumann, in "Why the Mind Is in the Head," 58.

108. Köhler, in "Why the Mind Is in the Head," 66–68.

109. Eccles, *The Neurophysiological Basis of Mind*, 345–346.

110. McCulloch, "Why the Mind Is in the Head," 99.

111. Ibid.

112. McCulloch in John von Neumann, "The General and Logical Theory of Automata," in *Cerebral Mechanisms in Behavior: The Hixon Symposium*, ed. Lloyd A. Jeffress (New York: Wiley, 1951), 32.

113. Charles Sanders Peirce, Collected Papers 5.145, *Volume V: Pragmatism and Pragmaticism*. In *Collected Papers of Charles Sanders Peirce*, ed. Charles Hartshorne and Paul Weiss (Cambridge, MA: Belknap Press/Harvard University Press, 1965), 90.

114. See Stathis Psillos, "An Explorer upon Untrodden Ground: Peirce on Abduction," in *Inductive Logic. Handbook of the History of Logic, Volume 10*, ed. Dov M. Gabbay, John Woods, and Stephan Hartmann (Amsterdam: Elsevier, 2011), 117–151.

115. Karl Lashley, "Functional Interpretation of Anatomic Patterns," in *Patterns of Organization in the Central Nervous System, Research Publications of the Association for Research in Nervous and Mental Diseases*, XXX (1952): 529–547, 541–542.

116. Warren S. McCulloch, "Introductory Remarks," in *Cybernetics: Causal and Feedback Mechanisms in Biological and Social Systems: Transactions of the Tenth Conference*, ed. Heinz von Foerster, Margaret Mead, and Hans Lukas Teuber (New York: Josiah Macy, Jr. Foundation, 1955), 15–18, 17. In Claus Pias (ed.) *Cybernetics—Kybernetik: The Macy Conferences 1946–1953* (Zürich-Berlin: Diaphanes, 2003), 687–688, 688.

117. Donald M. MacKay and Warren S. McCulloch, "The Limiting Informational Capacity of a Neuronal Link," *Bulletin of Mathematical Biophysics* 14 (1952): 127–135.

118. McCulloch, "Introductory Remarks," 17.

119. Warren S. McCulloch, "*Mysterium Iniquiatis* of Sinful Man Aspiring into the Place of God," *The Scientific Monthly* 80, no. 1 (1955): 35–39; in Warren S. McCulloch, *Embodiments of Mind*, 157–164 (Cambridge, MA: MIT Press, 1988), 163.

120. George A. Miller, "Review of *Cybernetics: Circular Causal and Feedback Mechanisms in Biological and Social Systems, Transactions of the Eighth Conference* by Heinz von Foerster, Margaret Mead, and Hans Lukas Teuber," *American Journal of Psychology* 66 No. 4 (October 1953): 661–663.

121. Miller, "Review of *Cybernetics*," 663.

122. Sabina Leonelli, "What Is in a Model? Combining Theoretical and Material Models to Develop Intelligible Theories," in *Modeling Biology: Structures, Behaviors, Evolution*, ed. Manfred D. Laubichler and G. M. Müller (Cambridge, MA: MIT Press, 2007), 15–36.

123. See also Richard Levins, "The Strategy of Model Building in Population Biology," *American Scientist* 54 (1966): 421–431.

124. David Resnik, "How-Possibly Explanations in Biology," *Acta Biotheoretica* 39, no. 2 (1991): 142–149. See also Robert N. Brandon, *Adaptation and Environment* (Princeton, NJ: Princeton University Press, 1990), 176–184.

125. Seymour Papert, "Introduction," in Warren S. McCulloch, *Embodiments of Mind* (Cambridge, MA: MIT Press, 1988), xxviii.

126. Norbert Wiener, *The Human Use of Human Beings* (Cambridge, MA: MIT Press, 1950), Chapters 1 and 2.

127. I use "interdiscipline" here in Galison's sense ("The Americanization of Unity," 66), as cross-disciplinary formulations arising out of the wartime context.

128. Hans Pols, "War Neurosis, Adjustment Problems in Veterans, and an Ill Nation: The Disciplinary Project of American Psychiatry during and after World War II," *Osiris* Second Series 22 (2007): 72–92; Hans Pols, "The Tunisian Campaign, War Neuroses, and the Reorientation of American Psychiatry during World War II," *Harvard Review of Psychiatry* 19 (2011): 313–320.

## Chapter 7

1. McCulloch to Wiesner, 31 May 1951, Folder "Wiesner, Jerome," Series I, WSM Papers, APS. Jerome Lettvin and Patrick D. Wall were also included in the invitation. Wiesner had been on staff at the Radiation Laboratory, and had then worked at Los Alamos in 1945 before returning to MIT in 1946 as Assistant Professor in Electrical Engineering. Wiesner eventually served as President John F. Kennedy's Special Assistant for Science and Technology and chaired the President's Science Advisory Committee.

2. Walter H. Pitts and Jerry Lettvin to Rosenblueth and Wiener, 14 November 1951, Box 10.144, NW Papers, MITASC.

3. Jerome B. Wiesner to Frank E. Mott, 18 December 1952, Folder "Teagle Foundation, Inc. #1," Series II, WSM Papers, APS.

4. Ibid.

5. I take my definition of "multidisciplinary" from Julie Thompson Klein, who uses the term to characterize practices where the disciplines involved are additive rather than integrative (Julie Thompson Klein, *Interdiscipliarity: History, Theory, and Practice* [Detroit: Wayne State University Press, 1990], 56–57). Jamie Cohen-Cole has made a similar argument for Harvard's Center for Cognitive Studies, arguing it was multidisciplinary not interdisciplinary (Jamie Cohen-Cole, "Instituting the Science of Mind: Intellectual Economies and Disciplinary Exchange at Harvard's Center for Cognitive Studies," *British Journal for the History of Science* 40, no. 4 [2007]: 567–597).

6. Warren S. McCulloch to Paul McCulloch, 26 May 1952, Folder "McCulloch, Paul" Series I, WSM Papers, APS. Paul was McCulloch's half-brother, born to his father's first wife Jane, who died soon after Paul's birth ("James W. McCulloch. Notes by daughter, Margaret C. McCulloch for David Sears McCulloch, Christmas, 1966," McCulloch Family Photographs, Box 5, Series VIII, WSM Papers, APS).

7. Warren S. McCulloch, Interview by Percy Saltzman, *The Day It Is*, Canadian Broadcasting Corporation, 1969.

8. Lewis Campbell and William Garnett, *The Life of James Clerk Maxwell, with Selections from his Correspondence and Occasional Writings*, Second Edition (London: Macmillan & Co., 1884), 16; Basil Mahon, *The Man Who Changed Everything: The Life of James Clerk Maxwell* (New York: Wiley, 2003), 3.

9. Warren S. McCulloch, Interview by Percy Saltzman, *The Day It Is*, Canadian Broadcasting Corporation, 1969.

10. Warren S. McCulloch, "Through the Den of the Metaphysician," lecture delivered to the Philosophical Club of the University of Virginia, 23 March 1948, English version of "Dans l'antre du métaphysicien," *Thales*, Volume 7 (Paris: Presses

Universitaires de France, 1951), 37–49. In McCulloch, *Embodiments of Mind*, 142–156, 143. Original quote in James Clerk Maxwell, "Address to the Mathematical and Physical Sections of the British Association," *British Association Report* Vol. XL, 15 September 1870.

11. Matthew Stanley, *Huxley's Church and Maxwell's Demon: From Theistic Science to Naturalistic Science* (Chicago: University of Chicago Press, 2015), 107.

12. Robert E. Kohler, *Lords of the Fly: Drosophila Genetics and the Experimental Life* (Chicago: University of Chicago Press, 1994), 6.

13. Manuel Blum, Interview with author, 4 June 2013.

14. Tara H. Abraham, "'The Materials of Science, the Ideas of Science, and the Poetry of Science': Warren McCulloch and Jerry Lettvin," *Interdisciplinary Science Reviews* 37, no. 3 (September 2012): 269–86.

15. Hunter Crowther-Heyck, "George A. Miller, Language, and the Computer Metaphor of Mind," *History of Psychology* 2 (1999): 37–64. For more on the institutional context of cognitive science during this period, see Jamie Cohen-Cole, *The Open Mind: Cold War Politics and the Sciences of Human Nature* (Chicago: University of Chicago Press, 2014), Chapter 6.

16. Margaret Boden, *Mind as Machine: A History of Cognitive Science*, 2 vols. (Oxford: Oxford University Press, 2006); Roberto Cordeschi, *The Discovery of the Artificial: Behavior, Mind, and Machines Before and Beyond Cybernetics* (Dordrecht: Kluwer, 2002); Pamela McCorduck, *Machines Who Think: A Personal Inquiry into the History and Prospects of Artificial Intelligence* (Natick, MA: AK Peters, 2004).

17. Roger Geiger, *Research and Relevant Knowledge: American Research Universities Since World War II* (Oxford: Oxford University Press, 1993).

18. For more on this see Paul Forman, "Behind Quantum Electronics: National Security as a Basis for Physical Research in the United States, 1940–1960," *Historical Studies in the Physical Sciences* 18, no. 1 (1987): 149–229; Nathan Reingold, "Science and Government in the United States since 1945," *History of Science* 32, no. 3 (1994): 361–386; Roger Geiger "Science, Universities, and National Defense, 1945–1970," *Osiris 2nd Ser.* Vol 7 (1992): 26–48; Stuart W. Leslie, "Science and Politics in Cold War America," in *The Politics of Western Science, 1964–1990*, ed. Margaret C. Jacob (Humanity Books, 1994), 199–233; Stuart W. Leslie, *The Cold War and American Science: The Military-Industrial-Academic Complex at MIT and Stanford* (Stanford, CA: Stanford University Press, 1993).

19. Hunter Heyck and David Kaiser, "Introduction," in "Focus: New Perspectives on Science and the Cold War," *Isis* 101, no.2, (2010): 362–366; Toby A. Appel, *Shaping Biology: The National Science Foundation and American Biological Research, 1945–1975* (Baltimore and London: Johns Hopkins University Press, 2000); Mark Solovey

"Science and the State during the Cold War: Blurred Boundaries and a Contested Legacy," *Social Studies of Science* 31, no. 2 (2001): 165–170; Mark Solovey, *Shaky Foundations: The Politics-Patronage-Social Science Nexus in Cold War America* (New Brunswick, NJ, and London: Rutgers University Press, 2013); Hunter Crowther-Heyck, "Patrons of the Revolution: Ideals and Institutions in Postwar Behavioral Science," *Isis* 97, no. 3 (2006): 420–46; Joel Isaac, "Introduction: The Human Sciences and Cold War America," *Journal of the History of the Behavioral Sciences* 47, no. 3 (2011): 225–231; Paul Erickson, "Mathematical Models, Rational Choice, and the Search for Cold War Culture," *Isis* 101, no.2 (2010): 386–392; Jamie Cohen-Cole "The Creative American: Cold War Salons, Social Science, and the Cure for Modern Society," *Isis* 100, no. 2 (2009): 219–62; Ron Robin, *Making the Cold War Enemy: Culture and Politics in the Military-Industrial-Academic Complex* (Princeton, NJ: Princeton University Press, 2011).

20. E.g., Joel Isaac "The Human Sciences in Cold War America," *The Historical Journal* 50, no. 3 (2007): 725–46; Angela N. H. Creager, *The Life of a Virus: Tobacco Mosaic Virus as an Experimental Model, 1930–1965* (Chicago: University of Chicago Press, 2002); Nicolas Rasmussen, "'Of 'Small Men', Big Science and Bigger Business: The Second World War and Biomedical Research in the United States," *Minerva: A Review of Science Learning and Policy* 40 (2002): 115–146.

21. David Engerman, "Rethinking the Cold War Universities: Some Recent Histories," *Journal of Cold War Studies* 5, no. 3 (Summer 2003): 80–95. See also Richard C. Lewontin, "The Cold War and the Transformation of the Academy," in Noam Chomsky, R. C. Lewontin, Ira Katznelson, Laura Nader, Richard Ohmann, David Montgomery, Immanuel Wallerstein, Ray Siever, and Howard Zinn, *The Cold War and the University: Toward and Intellectual History of the Postwar Years* (New York: New Press, 1997), 1–34.

22. See Roger L. Geiger, *Research and Relevant Knowledge*.

23. Noam Chomsky, "The Cold War and the University," in Noam Chomsky et al., *The Cold War and the University*, 171–194.

24. Roger L. Geiger, "Organized Research Units—Their Role in the Development of University Research," *Journal of Higher Education* 61, no. 1 (Jan.–Feb. 1990): 1–19; Geiger, *Research and Relevant Knowledge*, Chapter 2.

25. Ron Robin, *The Making of the Cold War Enemy*, 40.

26. "The Research Laboratory of Electronics at the Massachusetts Institute of Technology" December 1945, Box 4.18, Records of the RLE, MITASC.

27. Memorandum from Professor Hazen to President Compton, 10 March 1945, Box 10 Folder "Hazen, Harold, Admin Files", Records of the RLE, MITASC; "Memorandum of conversation February 3, 1945, between K.T. Compton, J.A. Stratton,

P.M. Morse, J.C. Slater" Box 10 Folder "Hazen, Harold Admin Files," Records of the RLE, MITASC.

28. Stuart W. Leslie, *The Cold War and American Science*, 20–32; Geiger, *Research and Relevant Knowledge*, 63.

29. Stuart W. Leslie, "Profit and Loss: The military and MIT in the Postwar Era," *Historical Studies in Physical and Biological Sciences* 21, no.1 (1990): 59–85, 65. The JSEP also helped create nine research centers across the country, all devoted to basic research in electronics. "Joint Services Electronics Program" Box 2 Folder 2, Records of the RLE, MITASC.

30. Leslie, *The Cold War and American Science*, 22–23; Julius A. Stratton, "RLE: The Beginning of an Idea," in *RLE: 1946+20* (Cambridge, MA: MIT Press, 1966), 1–6; 6.

31. "The Research Laboratory of Electronics at the Massachusetts Institute of Technology, December 1945" Box 4, Folder 18, Records of the RLE, MITASC.

32. Stuart Leslie, "Profit and Loss"; Daniel J. Kevles, "The National Science Foundation and the Debate over Postwar Research Policy, 1942–1945," *Isis* 68 (1977): 5–26.

33. As RLE Director Henry J. Zimmerman wrote in 1966, the "military value was especially apparent at the start of the Korean War, at which time many of the highly-skilled scientists in RLE applied their knowledge to classified military problems" (H.J. Zimmerman, "Programs and Accomplishments," 26 October 1966, Box 2 Folder 13, Records of the RLE, MITASC). See also Geiger, *Research and Relevant Knowledge*, 66–67.

34. David Kaiser, "Elephant on the Charles: Postwar Growing Pains," in *Becoming MIT: Moments of Decision*, ed. David Kaiser (Cambridge, MA: MIT Press, 2010), 103–121; 105.

35. Geiger, *Research and Relevant Knowledge*, 34.

36. Julius A. Stratton, "RLE—the Beginning of an Idea."

37. Harvey M. Saplosky, *Science and the Navy: The History of the Office of Naval Research* (Princeton: Princeton University Press, 1990), 4; Geiger, "Organized Research Units," 9–10.

38. Saplosky, *Science and the Navy*, 42–43. For the origins of the ONR, see Saplosky, *Science and the Navy*, Chapter 2.

39. Lewontin, "The Cold War and the Transformation of the Academy," 16.

40. Noam Chomsky recalls that Department of Defense funding during the 1950s was "free" in the kinds of projects it would support (Chomsky, "The Cold War University," 182–183).

41. Saplosky, *Science and the Navy*, 58.

42. For a sociological critique of the relation between state support and basic research, see Chandra Mukerji, *A Fragile Power: Scientists and the State*, (Princeton, NJ: Princeton University Press, 1989).

43. Saplosky, *Science and the Navy*, 57.

44. Jerome B. Wiesner, "The Communication Sciences—Those Early Days," in *Research Laboratory of Electronics, R.L.E.: 1946+20*, Research Laboratory of Electronics, Massachusetts Institute of Technology (Cambridge, MA, May 1966), 12–16; 12.

45. Jerome B. Wiesner, "The Communication Sciences—Those Early Days," 15.

46. Eventually, in 1963, a group including Marvin Minsky, John McCarthy, Walter Rosenblith, Robert Fano, and Herbert Teager was assembled by Albert Hill to discuss the design and construction of a time-sharing computer; this evolved into Project MAC, the first lab devoted to artificial intelligence, with funds from the Defense Advanced Research Projects Agency (DARPA).

47. "Research Groups Listed in R.L.E.—Q.P.R. at 5-Year Intervals," *Research Laboratory of Electronics, R.L.E.: 1946+20*, Research Laboratory of Electronics, Massachusetts Institute of Technology (Cambridge, MA, May 1966), 24–25.

48. Jerome Y. Lettvin, "Warren and Walter," in *The Collected Works of Warren S. McCulloch*, ed. Rook McCulloch (Salinas, CA: Intersystems, 1989), 514–529, 522. For more on McCulloch's roles as mentor during this period, and his relations with Lettvin in particular, see Tara H. Abraham, "'The Materials of Science, the Ideas of Science, and the Poetry of Science': Warren McCulloch and Jerry Lettvin," *Interdisciplinary Science Reviews* 37, no. 3 (September 2012): 269–86.

49. McCulloch to Members of the Neurological Study Section, National Institute of Health, 26 February 1958, Folder "National Science Foundation #2," Series II, WSM Papers, APS.

50. Peter Greene, "Warren McCulloch's Advice" 1972/1974, Folder "Peter Greene" Series V, WSM Papers, APS.

51. "Report for the Teagle Foundation, Incorporated, 1958," Folder "Teagle Foundation, Inc. #2," Series II, WSM Papers, APS.

52. McCulloch to Frank E. Mott, 5 December 1952, Folder "Teagle Foundation, Inc. #1," Series II, WSM Papers, APS.

53. Memo from Henry J. Zimmerman to Jerome Y. Lettvin, 6 December 1966, Box 1.1, JYL Papers, MITASC.

54. Jerome Y. Lettvin, "Jerome Y. Lettvin," in *Talking Nets: An Oral History of Neural Networks*, ed. James A. Anderson and Edward Rosenfeld (Cambridge, MA: MIT Press, 1998), 1–21; 12.

55. Manuel Blum, Interview with author, 28 May 2013.

56. McCulloch to Members of the Neurological Study Section, National Institute of Health, 26 February 1958, Folder "National Science Foundation #2," Series II, WSM Papers, APS.

57. Edward Jones-Imhotep, "Maintaining Humans," in *Cold War Social Science: Knowledge Production, Liberal Democracy, and Human Nature*, ed. Mark Solovey and Hamilton Cravens (New York: Palgrave Macmillan, 2012), 175–195.

58. Jones-Imhotep, "Maintaining Humans," 178–179.

59. E.g., Warren S. McCulloch, "The Reliability of Biological Systems," in *Self-Organizing Systems*, ed. M. C. Yovitz and S. Cameron (Pergamon Press, 1960), 264–280. In *The Collected Works of Warren S. McCulloch, Volume 4*, ed. Rook McCulloch (Salinas, CA: Intersystems, 1989), 1193–1210. Warren S. McCulloch, Michael A. Arbib, and Jack Cowan, "Neurological Models and Integrative Processes," in *Self-Organizing Systems*, ed. M. C. Yovitz, G. T. Jacobi, and D. G. Goldstein (Washington: Spartan Books, 1962), 49–59. In *The Collected Works of Warren S. McCulloch, Volume 4*, ed. Rook McCulloch (Salinas, CA: Intersystems, 1989), 1260–1270.

60. John von Neumann, *The Computer and the Brain*, 2nd ed. (New Haven and London: Yale University Press, 1958/2000).

61. Boden, *Mind as Machine*, 890.

62. E.g., Warren S. McCulloch, "Where is Fancy Bred?" in *Lectures on Experimental Psychiatry*, ed. Henry W. Brosin (Pittsburgh, PA: University of Pittsburgh Press, 1961), 311–324. In Warren S. McCulloch, *Embodiments of Mind* (Cambridge, MA: MIT Press, 1988), 216–229; Warren S. McCulloch "Agathe Tyche of Nervous Nets: The Lucky Reckoners," in *Embodiments of Mind* (Cambridge, MA: MIT Press, 1988).

63. Warren S. McCulloch, "Reliable Systems Using Unreliable Units," in *Disorders of Communication* (Volume XLII, Research Publications of the Association for Research in Nervous and Mental Disease), ed. David McKenzie Rioch and Edwin A. Weinstein (Baltimore, MD: Williams and Wilkins, 1964), 19–28. In *The Collected Works of Warren S. McCulloch, Volume 4*, ed. Rook McCulloch (Salinas, CA: Intersystems, 1989), 1292–1301.

64. McCulloch to Members of the Neurological Study Section, National Institute of Health, 26 February 1958, Folder "National Science Foundation #2," Series II, WSM Papers, APS.

65. G. Moruzzi and H. W. Magoun, "Brain Stem Reticular Formation and Activation of the EEG," *Electroencephalography and Clinical Neurophysiology* 1 (1949): 455–73.

66. J. Droogleever-Fortuyn, "Reticular Formation and Choice of Behavior," in *The Collected Works of Warren S. McCulloch, Volume 4*, ed. Rook McCulloch (Salinas, CA: Intersystems, 1989), 1302–1305; 1302.

67. E.g., Warren S. McCulloch and William L. Kilmer, "The Command and Control System of the Vertebrates," in *Proceedings of the International Federation for Information Processing, Congress, New York (May 24–29, 1965), Volume 2*, ed. Wayne A. Kalenich (Washington: Spartan Books, 1966), 636. In *The Collected Works of Warren S. McCulloch, Volume 4*, ed. Rook McCulloch (Salinas, CA: Intersystems, 1989), 1306; William Kilmer and Warren S. McCulloch, "The Reticular Formation Command and Control System," in *Information Processing in the Nervous System* (Proceedings of the Symposium on Information Processing in the Nervous System, State University of NY at Buffalo, 21–24 October 1969), ed. K. N. Leibovic (Springer-Verlag, 1969), 99–102. In *The Collected Works of Warren S. McCulloch, Volume 4*, ed. Rook McCulloch (Salinas, CA: Intersystems, 1989), 1322–1332.

68. William L. Kilmer, Warren S. McCulloch, and J. Blum, with E. Craighill and D. Peterson, "Towards a Theory of the Reticular Formation," in *The Mind: Biological Approaches to its Functions*, ed. William C. Corning and Martin Balaban, 177–232 (New York: Intersciences, 1968).

69. Warren S. McCulloch, Report to Dr. Harvey E. Savely, Re: Special Orders Number A-633, Folder United States Air Force Office of Scientific Research, Series II, WSM Papers, APS.

70. Warren S. McCulloch to David McKenzie Rioch, 28 February 1951, Folder United States Army Medical Service Graduate School, Medical Sciences Basic Course, Series II, WSM Papers, APS.

71. E.g., Justin Leiber, *An Invitation to Cognitive Science* (Cambridge, MA: Blackwell, 1991); Barbara von Eckhardt, *What Is Cognitive Science?* (Cambridge, MA: MIT Press, 1993); Howard Gardner, *The Mind's New Science: A History of the Cognitive Revolution* (New York: Basic Books, 1985); Margaret A. Boden, *Mind as Machine: A History of Cognitive Science* (Oxford: Oxford University Press, 2006). For the history of AI. see Pamela McCorduck, *Machines Who Think*; Daniel Crevier, *AI: The Tumultuous History of the Search for Artificial Intelligence* (Basic Books, 1993).

72. From Paul N. Edwards, *The Closed World: Computers and the Politics of Discourse in Cold War America* (Cambridge, MA: MIT Press, 1996), 178–9.

73. Jerome S. Bruner, Jacqueline J. Goodnow, and George A. Austin, *A Study of Thinking* (New York: Wiley, 1956); George A. Miller, E. Galanter, and Karl H. Pribram, *Plans and the Structure of Behavior* (New York: Holt, 1960).

74. Hunter Crowther-Heyck, "George A. Miller, Language, and the Computer Metaphor of Mind," 57n3.

75. E.g., George A. Miller, "The Magical Number Seven, Plus or Minus Two: Some Limits on Our Capacity for Processing Information," *Psychological Review* 63, no. 2 (1956): 81–97; Cohen-Cole, "Instituting the Science of Mind," 573–4.

76. On Herbert Simon, see Crowther-Heyck, *Herbert A. Simon*.

77. J. McCarthy, M. L. Minsky, N. Rochester, and C. E. Shannon, "A Proposal for the Dartmouth Summer Research Project on Artificial Intelligence," 31 August 1955 (http://www-formal.stanford.edu/jmc/history/dartmouth/dartmouth.html; accessed 11 July 2013).

78. John McCarthy, Review of Brian P. Boomfield (ed.) *The Question of Artificial Intelligence: Philosophical and Sociological Perspectives* (London, New York, Sydney: Croom Helm, 1987) (http://www-formal.stanford.edu/jmc/reviews/bloomfield/bloomfield.html, accessed 11 July 2013).

79. Ronald R. Kline, "Cybernetics, Automata Studies, and the Dartmouth Conference on Artificial Intelligence," *IEEE Annals of the History of Computing* (Oct/Dec 2011): 5–16.

80. Edwards, *The Closed World*, 253.

81. Heinz von Foerster to Warren S. McCulloch, 31 March 1958, Folder "Von Foerster, H.," Series I, WSM Papers, APS.

82. Edward A. Feigenbaum and Julian Feldman, "Preface," in Edward A. Feigenbaum and Julian Feldman (eds.), *Computers and Thought* (New York: McGraw-Hill, 1963), v–vii. McCulloch was in the latter camp: he received support from the "Information Systems Branch" starting in 1962 for a project on "Neurophysiological Self-Organizing Systems," which seems to have been renewed until at least 1965. Dorothy M. Gilford to R. P. Webber, Office of the Director, MIT, 4 January 1965; Richard H. Wilcox to Warren S. McCulloch, 30 April 1965, Folder "Office of Naval Research #2," Series II, WSM Papers, APS. He also presented his work, along with Manuel Blum, at a meeting in 1961 on Self-Organizing Systems, sponsored by the ONR (Heinz von Foerster and George W. Zopf, Jr. (eds.) *Principles of Self-Organization: Transactions of the University of Illinois Symposium on Self-Organization, Robert Allerton Park, 8 and 9 June, 1961* (New York: Pergamon Press, 1962).

83. Marvin Minsky, "Warren McCulloch: For Whom the World Was a Stage." Interview with Christopher Sykes, 29–31 January 2011, Web of Stories, http://www.webofstories.com/play/marvin.minsky/22, accessed 10 July 2013.

84. Marvin Minsky, "Warren McCulloch: For Whom the World Was a Stage."

85. Lenore Blum, Interview with author, 28 May 2013.

86. Alex Andrew, "Tribute to the Life and Work of Rook McCulloch," *Kybernetes* 22, no. 3 (1993), 4.

87. For more on McCulloch's mentoring practices, see Abraham, "'The Materials of Science, the Ideas of Science, and the Poetry of Science'."

88. "Information Sciences Summary, 'Neurophysiological Self-Organizing Systems,' n.d., for Principal Investigatory Warren S. McCulloch, Project Number NR

049–164/3–12–65; Grant Number Nonr(G)-00063–65," Folder "United States Nave. Office of Naval Research. Travel #1," Series II, WSM Papers, APS.

89. Blum, a Venezuelan-born mathematician, was educated at MIT for B.S. and M.S. in electrical engineering (1961) and received his Ph.D. in mathematics under the supervision of Marvin Minsky in 1964; he was research assistant in the Neurophysiology Group at MIT from 1959 to 1964.

90. Michael Arbib began with a summer appointment in the RLE with McCulloch in 1961, and was reappointed as a research assistant through to summer 1962: Ralph A. Sayers to Arbib, 4 April 1961; Richard D. Schafer to Arbib, 15 May 1961 (Folder "Arbib, Michael," Series I, WSM Papers, APS). After leaving the RLE, to eventually take up a position at the School of Mathematics at the University of New South Wales, Australia, Arbib often wrote affectionate postcards and letters to McCulloch and "the Gang." McCulloch wrote reference letters for Arbib.

91. Jack Cowan began working with McCulloch as early as 1960, when he was working on his Master's degree in Engineering. McCulloch was thrilled with the work Cowan was doing and write to Colin Cherry of Imperial College in London, hoping he would return to England to complete a Ph.D., and promising he would "stir up" whatever funds were needed in order to ensure this could happen (McCulloch to Colin Cherry, 4 May 1960, Folder "Cowan, Jack," Series I, WSM Papers, APS). Originally Cowan's presence at the RLE was made possible by McCulloch's NIH grant. McCulloch did try secure US federal funding for Cowan (Cowan to McCulloch, 12 November 1962; McCulloch to James B. Hargering, 15 November 1962, Folder "Cowan, Jack," Series I, WSM Papers, APS).

92. Manuel Blum, Interview with author, 14 June 2013.

93. Manuel Blum, Interview with author, 28 May 2013.

94. Manuel Blum, Interview with author, 30 May 2013.

95. Laura Otis, *Müller's Lab* (New York: Oxford University Press, 2007).

96. For example, McCulloch attended the Fifth Annual Meeting of the Academy of Psychosomatic Medicine in 1958 (William S. Kroger to McCulloch, 20 January 1958, Folder "Academy of Psychosomatic Medicine," Series I, WSM Papers, APS).

97. McCulloch to E. J. Alexander, 28 September 1953, Folder "Alexander, E. J.," Series I, WSM Papers, APS.

98. McCulloch to Alexander, 28 September 1953.

99. E. J. Alexander to McCulloch, 1 October 1953, Folder "Alexander, E. J.," Series I: Correspondence, WSM Papers, APS.

100. Warren S. McCulloch, "Introductory Remarks," at the Tenth Macy Meeting on Cybernetics (March 1953), in *Cybernetics—Kybernetik: The Macy Conferences,*

*1946–1953*, ed. Claus Pias (Zürich-Berlin: Diaphanes, 2003), 687–688; Warren S. McCulloch, "Appendix I: Summary of the Points of Agreement Reached in the Previous Nine Conferences on Cybernetics," at the Tenth Macy Meeting on Cybernetics, in in *Cybernetics—Kybernetik: The Macy Conferences, 1946–1953*, ed. Claus Pias (Zürich-Berlin: Diaphanes, 2003), 719–725; Warren S. McCulloch, "What Is a Number, That a Man May Know It, and a Man, That He May Know a Number?" *Embodiments of Mind*, 1–18. (Cambridge, MA: MIT Press, 1988). (Originally published as The Ninth Annual Alfred Korzybski Memorial Lecture, *General Semantics Bulletin Nos. 26 and 27*, Lakeville, CT: Institute of General Semantics, 1961, pp. 7–18.)

101. E.g., Warren S. McCulloch, "Where Is Fancy Bred?" in *Lectures on Experimental Psychiatry*, ed. Henry W. Brosin (Pittsburgh, PA: University of Pittsburgh Press, 1961), 311–324. In Warren S. McCulloch, *Embodiments of Mind* (Cambridge, MA: MIT Press, 1988), 216–229. This is a reference to a song in Shakespeare's *The Merchant of Venice*, Act III, Scene 2: "Tell me where is fancy bred, Or in the heart or in the head? ..."

102. Warren S. McCulloch, "Abracadabra," in *Mental Patients in Transition*, ed. M. Greenblatt, D. J. Levinson, and G. J. Klerman (Springfield, IL: Charles C. Thomas, 1961), in *The Collected Works of Warren S. McCulloch, Volume 4*, ed. Rook McCulloch (Salinas, CA: Intersystems, 1989), 1244–1255.

103. Samuel Johnson, *Rasselas, Prince of Abissinia: A Tale* (London: E. and S. Harding, Pall Mall, 1796), Chapter X. Eighteenth Century Collections Online.

104. Warren S. McCulloch, *The Natural Fit* (Chicago Literary Club, 1959), 23.

105. Oxford English Dictionary.

106. McCulloch, *The Natural Fit*, 5.

107. McCulloch, *The Natural Fit*, 44.

108. Heinrich Klüver to Warren S. McCulloch, 20 December 1959, Folder "Poetry Letters," Series II, WSM Papers, APS.

109. John F. Fulton to Warren S. McCulloch, 22 December 1959, Folder "Poetry Letters," Series II, WSM Papers, APS.

110. W. Ross Ashby, Review of Warren S. McCulloch, *Embodiments of Mind* (Cambridge, MA: MIT Press, 1965), *Journal of Nervous and Mental Disease* 142, no. 4 (April 1966), 491.

111. Kenneth E. Boulding, *Scientific American* (April 1966), 138.

112. W. Grey Walter, Review of Warren S. McCulloch, *Embodiments of Mind* (Cambridge, MA: MIT Press, 1965); *Electroencephalography and Clinical Neurophysiology* 21 (1966): 311–312.

113. Walter, 312.

114. Tara H. Abraham. "'The Materials of Science, the Ideas of Science, and the Poetry of Science': Warren McCulloch and Jerry Lettvin," *Interdisciplinary Science Reviews* 37, no. 3 (2012): 269–86.

115. Karl Pribram, "Brilliant, Lively, Productive, and Wrong," *PsycCritiques* 11, no. 12 (December 1966): 580–581.

## Chapter 8

1. Judith Butler, *Gender Trouble: Feminism and the Subversion of Identity* (New York: Routledge, 1990), 24, cited in Rebecca Herzig, "On Performance, Productivity, and Vocabularies of Motive in Recent Studies of Science," *Feminist Theory* (2004) 5: 127–147.

2. Clifford Geertz, "The Way We Think Now: Towards an Ethnography of Modern Thought," in Clifford Geertz, *Local Knowledge: Further Essays in Interpretative Anthropology* (Basic Books, 2000), 155.

3. Roger Smith, "Review of Kenton Kroker, *The Sleep of Others and the Transformation of Sleep Research*," *History of the Human Sciences* 22, no. 5 (2008): 108.

4. E.g., Paul Forman, "On the Historical Forms of Knowledge Production and Curation: Modernity Entailed Disciplinarity, Postmodernity Entails Antidisciplinarity," in *Clio Meets Science: The Challenges of History*, ed. Robert E. Kohler and Kathryn M. Olesko, *Osiris* 27 (2012): 56–97; Ananta Kumar Giri, "Transcending Disciplinary Boundaries: Creative Experiments and the Critiques of Modernity," *Critique of Anthropology* 18, no. 4 (1998): 379–404; 382.

5. James M. M. Good, "Disciplining Social Psychology: A Case Study of Boundary Relations in the History of the Human Sciences," *Journal of the History of the Behavioral Sciences* 36, no. 4 (Fall 2000): 383–403; 385.

6. Thomas Gieryn, "Boundary Work and the Demarcation of Science from Non-Science: Strains and Interests in Professional Ideologies of Scientists," *American Sociological Review* 48 (1983): 781–795; Julie Thompson Klein, "Blurring, Cracking, and Crossing: Permeation and the Fracturing of Discipline," in *Knowledges: Historical and Critical Studies in Disciplinarity*, ed. Ellen Messer-Davidow, David R. Shumnay, and David J. Sylvan (Charlottesville: University of Virginia Press, 1993), 185–211; Ellen Messer-Davidow, David R. Shumnay, and David J. Sylvan, "Disciplinary Ways of Knowing," in *Knowledges: Historical and Critical Studies in Disciplinarity*, ed. Ellen Messer-Davidow, David R. Shumnay, and David J. Sylvan (Charlottesville: University of Virginia Press, 1993), 1–21, 4.

7. Ian Hacking, "The Disunities of the Sciences," in *The Disunity of Science: Boundaries, Contexts, and Power*, ed. Peter Galison and David J. Stump (Stanford, CA: Stanford University Press, 1996), 37–74; 43–52.

8. Roger Smith, *The Human Sciences* (New York: W.W. Norton, 1997), 801. For more on the context of Cold War human sciences, see Paul Erickson, Judy L. Klein, Lorraine Daston, Rebecca Lemov, Thomas Sturm, and Michael D. Gordin, *How Reason Almost Lost Its Mind: The Strange Career of Cold War Rationality* (Chicago: University of Chicago Press, 2013).

9. Margaret Boden uses the term revolution reservedly, and the gradual emergence of the cognitive sciences as a unified field in Boden's view is closely linked to a critique and overthrow of the assumptions of behaviorism in psychology (Boden, *Mind as Machine*, 238–241). For a broad analysis of this process, see Hunter Crowther-Heyck, *Herbert A. Simon: The Bounds of Reason in Modern America* (Baltimore, MD: Johns Hopkins University Press, 2005), Chapter 11.

10. E.g., Allen Newell, J. C. Shaw, and Herbert A. Simon, "Elements of a Theory of Human Problem Solving," *Psychological Review* 65 (1958): 151–66; Marvin Minsky, "Steps toward Artificial Intelligence," *Proceedings of the Institute for Radio Engineers* 49 (1961): 8–30.

11. For more details on this intellectual landscape, see Boden, *Mind as Machine*, *Volume II*, Chapters 12 and 14.

12. T. J. Sejnowski, C. Koch, and P. S. Churchland, "Computational Neuroscience," *Science* 241 (1988): 1299–1306.

13. Boden, *Mind as Machine*, 1114.

14. See e.g., Stephen Grossberg, "The Attentive Brain," *American Scientist* 83 (1995): 438–449.

15. Boden, *Mind as Machine*, Chapter 12.

16. Ian Hacking, "The Disunities of the Sciences," in *The Disunity of Science: Boundaries, Contexts, and Power*, ed. Peter Galison and David J. Stump (Stanford, CA: Stanford University Press, 1996), 37–74. Modeling has become a key feature of knowledge production in other areas of inquiry: on the social sciences, see Hunter Heyck, *The Age of System: Understanding the Development of Modern Social Science.* (Baltimore, MD: Johns Hopkins University Press, 2015), Chapter 6.

17. C. G. Gross, quoted in Boden, *Mind as Machine*, 1153.

18. Boden, *Mind as Machine*, 1167.

19. Andrew Scull, "A Psychiatric Revolution," *The Lancet* 275, no. 9722 (10 April 2010), 1246–1247. Scull argues that the emergence and wide acceptance of the Diagnostic and Statistical Manual of Mental Disorders and the concomitant rise of psycho-pharmaceuticals had much to do with the eventual decline of psychoanalysis.

20. Lenore Blum, Interview with author, 28 May 2013.

21. David McCulloch, Personal Communciation, 4 February 2015.

22. Percy Saltzman, *The Day It Is*, Canadian Broadcasting Corporation, 1969.

23. Ralph W. Gerard, "Warren Sturgis McCulloch: Rebel Genius," *Transactions of the American Neurological Association* 95 (1970): 344–345.

24. Anthony Giddens, *Modernity and Self-Identity: Self and Society in the Late Modern Age* (Stanford, CA: Stanford University Press, 1991), 2.

25. Karl Pribram, "Brilliant, Lively, Productive, and Wrong," *PsycCritiques* 11, no. 12 (December 1966): 580–581.

# References

## Unpublished Sources and Archives

American Philosophical Society, Philadelphia [APS]
  Warren S. McCulloch Papers (BM139) (WSM Papers)

Library of Congress, Washington, DC
  John von Neumann Papers

Massachusetts Institute of Technology Archives and Special Collections [MITASC]
  Jerome Y. Lettvin Papers (MC525) (JYL Papers)
  Norbert Wiener Papers (MC22) (NW Papers)
  Records of the Research Laboratory of Electronics (AC186) (Records of the RLE)

Rockefeller Archive Center, Pocantico Hills, NY [RAC]
University of Chicago Department of Special Collections [UCDSC]
University of Illinois at Chicago Archives [UICA]
University of Illinois at Urbana-Champaign Archives [UIUCA]
  Arthur Cutts Willard Papers

Yale University Archives [YUA]
  Anson Phelps-Stokes Family Papers
  Institute of Human Relations
  School of Medicine, Records of the Dean

## Interviews

Lenore Blum
Manuel Blum
Taffy Holland
Jerome Y. Lettvin
David McCulloch
Mary Jean Vasiloff

## Published Sources

"III.—The Medical School of Columbia University: College of Physicians and Surgeons," *The Lancet* 213 (5 January 1929), e9–e11.

Abir-Am, Pnina. 1987. "The Biotheoretical Gathering, Transdisciplinary Authority, and the Incipient Legitimation of Molecular Biology in the 1930s: New Perspective on the Historical Sociology of Science." *History of Science* 25 (67): 1–70.

Abraham, Tara H. 2012. "'The Materials of Science, the Ideas of Science, and the Poetry of Science': Warren McCulloch and Jerry Lettvin." *Interdisciplinary Science Reviews* 37 (3): 269–286.

Abraham, Tara H. 2012. "Transcending Disciplines: Scientific Styles in Studies of the Brain in Mid-Twentieth-Century America." *Studies in History and Philosophy of Biological and Biomedical Sciences* 43:552–568.

Abraham, Tara H. 2004. "Nicolas Rashevsky's Mathematical Biophysics." *Journal of the History of Biology* 37:333–385.

Abraham, Tara H. 2003. "From Theory to Data: Representing Neurons in the 1940s." *Biology & Philosophy* 18 (3): 415–426.

Abraham, Tara H. 2002. "(Physio)logical Circuits: The Intellectual Origins of the McCulloch–Pitts Neural Networks." *Journal of the History of the Behavioral Sciences* 38 (1): 3–25.

Adrian, Edgar D. 1947. *The Physical Background of Perception.* Oxford: Clarendon Press.

Adrian, Edgar D. 1914. "The All-or-None Principle in Nerve." *Journal of Physiology* 47:460–474.

Adrian, Edgar D. 1912. "On the Conduction of Subnormal Disturbances in Normal Nerve." *Journal of Physiology* 45:389–412.

Adrian, Edgar D., Frederic Bremer, and Herbert H. Jasper, eds. 1954. *Brain Mechanisms and Consciousness.* Oxford: Blackwell.

Adrian, Edgar D., and Yngve Zotterman. 1926. "The Impulses Produced by Sensory Nerve, Part 2: The Response of a Single End-Organ." *Journal of Physiology* 61:151–171.

Aizawa, Kenneth. September 2012. "Warren McCulloch's Turn to Cybernetics: What Walter Pitts Contributed." *Interdisciplinary Science Reviews* 37 (3): 206–217.

Aizawa, Kenneth, and Mark Schlatter. 2007. "Walter Pitts and 'A Logical Calculus.'" *Synthese* 162:235–250.

"Albert Poffenberger, 92; Professor at Columbia," *New York Times*, December 27, 1977, 38.

Allen, Frederick Lewis. 1951. *Only Yesterday: An Informal History of the Nineteen-Twenties*. New York: Harper.

Alten, Diana. Spring 1985. "Rufus Jones and the American Friend: A Quest for Unity." *Quaker History* 74 (1): 41–48.

Andrew, Alex. 1993. "Tribute to the Life and Work of Rook McCulloch." *Kybernetes* 22 (3): 4.

Anonymous. November 27, 1937. "Huxley, the Brain, and the Mind: Professor Adrian's Huxley Lecture." *British Medical Journal* 1082–1083.

Appel, Toby. 2000. *Shaping Biology: The National Science Foundation and the Shaping of American Biological Research, 1945–1975*. Baltimore, MD: Johns Hopkins University Press.

Arbib, Michael A. 2000. "Warren McCulloch's Search for the Logic of the Nervous System." *Perspectives in Biology and Medicine* 43 (2): 193–216.

Asaro, Peter. 2006. "Working Models and the Synthetic Method: Electronic Brains as Mediators between Neurons and Behavior." *Science Studies* 19 (1): 12–34.

Ashby, W. Ross. 1966. "Review of Warren S. McCulloch, *Embodiments of Mind* (Cambridge, MA: MIT Press, 1965)." *Journal of Nervous and Mental Disease* 142 (4): 491.

Aubin, David, and Charlotte Bigg. 2007. "Neither Genius nor Context Incarnate: Norman Lockyer, Jules Janssen, and the Astrophysical Self." In *The History and Poetics of Scientific Biography*, ed. Thomas Söderqvist, 51–70. Aldershot: Ashgate.

Baars, Bernard J. 1986. *The Cognitive Revolution in Psychology*. New York: Guilford Press.

Bailey, Percival, Gerhardt von Bonin, Hugh W. Garol, Warren S. McCulloch, E. Roseman, and Anibal Silvera. 1944. "Functional Organization of the Medial Aspect of the Primate Cortex." *Journal of Neurophysiology* 7:51–55.

Bailey, Percival, Johannes G. Dusser de Barenne, Hugh W. Garol, and Warren S. McCulloch. 1940. "Sensory Cortex of Chimpanzee." *Journal of Neurophysiology* 3:469–485.

Barad, Karen. 2007. *Meeting the Universe Halfway: Quantum Physics and the Entanglement of Matter and Meaning.* Duke University Press.

Barad, Karen. 2003. "Posthumanist Performativity: Toward an Understanding of How Matter Comes to Matter." *Signs (Chicago, IL)* 28 (3): 801–832.

Barry, Andrew, and Georgina Born. 2013. "Interdisciplinarity: Reconfigurations of the Social and Natural Sciences." In *Interdisciplinarity: Reconfigurations of the Social and Natural Sciences*, ed. Andrew Barry and Georgina Born, 1–56. London: Routledge.

Bartholomay, A. F., George Karreman, and Herbert D. Landahl. 1972. "Nicolas Rashevsky." *Bulletin of Mathematical Biophysics* 34 [no pagination].

Bartley, Samuel H., and George H. Bishop. 1933. "Factors Determining the Form of the Electrical Response from the Optic Cortex of the Rabbit." *American Journal of Physiology* 103:173–184.

Benjafield, John G. 2010. "The Golden Section and American Psychology, 1892–1938." *Journal of the History of the Behavioral Sciences* 46 (1): 52–71.

Bentley, Madison, and E. V. Cowdry, eds. 1934. *The Problem of Mental Disorder: A Study Undertaken by the Committee on Psychiatric Investigations, National Research Council.* New York: McGraw-Hill.

Besch, Michael D. 2002. *A Navy Second to None: The History of U.S. Naval Training in World War I.* Westport, CT: Greenwood Press.

Biagioli, Mario. 1993. *Galileo Courtier: The Practice of Science in an Age of Absolutism.* Chicago: University of Chicago Press.

Blaisdell, Russell E. 1931. "The Rockland State Hospital." *Psychiatric Quarterly* 5 (1): 193–195.

Blustein, Bonnie Ellen. 1993. "Medicine as Biology: Neuropsychiatry at the University of Chicago, 1928–1939." *Perspectives on Science* 1 (3): 416–444.

Blustein, Bonnie Ellen. 1992. "Percival Bailey and Neurology at the University of Chicago, 1928–1939." *Bulletin of the History of Medicine* 66:90–113.

Blustein, Bonnie Ellen. 1981. "'A Hollow Square of Psychological Science': American Neurologists and Psychiatrists in Conflict." In *Madhouses, Mad-Doctors, and Madmen*, ed. Andrew Scull, 241–270. Philadelphia: University of Pennsylvania Press.

Blustein, Bonnie Ellen. 1979. "New York Neurologists and the Specialization of American Medicine." *Bulletin of the History of Medicine* 53 (2): 170–183.

Boden, Margaret. 2006. *Mind as Machine: A History of Cognitive Science*, 2 vols. Oxford: Oxford University Press.

Boole, George. 1854. *An Investigation of the Laws of Thought*. London.

Borck, Cornelius. 2006. "Between Local Cultures and National Styles: Units of Analysis in the History of Electroencephalography." *Comptes Rendus de l'Académie des Sciences,* série Biologies 329: 450–459.

Borck, Cornelius. 2005. *Hirnströme: Eine Kulturgeschichte Der Elektroencephalographie*. Wallstein Verlag.

Bordogna, Francesca. 2008. *William James at the Boundaries: Philosophy, Science, and the Geography of Knowledge*. Chicago: University of Chicago Press.

Boring, Edwin G. 1929. *A History of Experimental Psychology*. New York: Century Co.

Boulding, Kenneth. 1966. "[Review of] *Endowments of Mind*." *Scientific American,* April: 138.

Bowker, Geof. 1993. "How to Be Universal: Some Cybernetic Strategies, 1943–1970." *Social Studies of Science* 23:107–127.

Braithwaite, Richard B. 1964. "Causal and Teleological Explanation." In *Scientific Explanation: A Study of the Function of Theory, Probability, and Law in Science* (Based upon the Tanner Lectures, 1946), 319–341. Cambridge: Cambridge University Press.

Brand, Jeanne. 1980. "Neurology and Psychiatry." In *The Education of American Physicians: Historical Essays*, ed. Ronald L. Numbers, 226–249. Berkeley: University of California Press.

Brandon, Robert N. 1990. *Adaptation and Environment*. Princeton, NJ: Princeton University Press.

Braslow, Joel T. 1997. *Mental Ills and Bodily Cures: Psychiatric Treatment in the First Half of the Twentieth Century*. Berkeley, CA: University of California Press.

Bromberg, Walter. 1982. *Psychiatry Between the Wars, 1918–1945: A Recollection*. Westport, CT: Greenwood Press.

Brown, Elspeth H. 2005. *The Corporate Eye: Photography and the Rationalization of American Commercial Culture*. Baltimore, MD: Johns Hopkins University Press.

Brown, Theodore M. 1987. "Alan Gregg and the Rockefeller Foundation's Support of Franz Alexander's Psychosomatic Research." *Bulletin of the History of Medicine* 61 (2): 155–182.

Bruner, Jerome S., Jacqueline J. Goodnow, and George A. Austin. 1956. *A Study of Thinking*. New York: Wiley.

Burnham, John C., ed. 2012. *After Freud Left: A Century of Psychoanalysis in America*. Chicago: University of Chicago Press.

Burnham, John C. 1987. *How Superstition Won and Science Lost: Popularizing Science and Health in the United States*. New Brunswick, NJ: Rutgers University Press.

Burnham, John C. 1979. "From Avant-Garde to Specialism: Psychoanalysis in America." *Journal of the History of the Behavioral Sciences* 15:128–134.

Burrow, Gerard N. 2002. *A History of Yale's School of Medicine: Passing Torches to Others*. New Haven, CT: Yale University Press.

Butler, Judith. 1990. *Gender Trouble: Feminism and the Subversion of Identity*. New York: Routledge.

Campbell, Lewis, and William Garnett. 1884. *The Life of James Clerk Maxwell, with Selections from his Correspondence and Occasional Writings*. 2nd ed. London: Macmillan & Co.

Capshew, James H. 1999. *Psychologists on the March: Science, Practice, and Professional Identity in America, 1929–1969*. Cambridge: Cambridge University Press.

Carnap, Rudolf. 1928. *Der Logische Aufbau der Welt*. Berlin: Weltkreis-Verlag.

Carter, Richard. 1999. "William Carlos Williams (1883–1963): Physician-Writer and 'Godfather of Avant-Garde Poetry.'" *Annals of Thoracic Surgery* 67:1512–1517.

Casamajor, L. 1943. "Notes for an Intimate History of Neurology and Psychiatry." *American Journal of Nervous and Mental Disease* 98:600–608.

Casper, Stephen T. March 2014. "History and Neuroscience: An Integrative Legacy." *Isis* 105 (1): 123–132.

Casper, Stephen T. 2010. "A Revisionist History of American Neurology." *Brain* 133:638–642.

Casper, Stephen T. 2008. "Conjunctures in Anglo-American Neurology: Lewis H. Weed and Johns Hopkins Neurology, 1917–1942." *Bulletin of the History of Medicine* 82:646–671.

Ceccarelli, Leah. 2001. *Shaping Science with Rhetoric: The Cases of Dobzhansky, Schrödinger, and Wilson*. Chicago: University of Chicago Press.

Ceccarelli, Leah. 1998. "Polysemy: Multiple Meanings in Rhetorical Criticism." *Quarterly Journal of Speech* 84 (4): 395–415.

Changeux, Jean-Pierre. 2004. *Physiology of Truth: Neuroscience and Human Knowledge*. Trans. M. B. DeBevoise. Cambridge, MA: Harvard University Press.

Chomsky, Noam. 1997. "The Cold War and the University." In Noam Chomsky, R. C. Lewontin, Ira Katznelson, Laura Nader, Richard Ohmann, David Montgomery, Immanuel Wallerstein, Ray Siever, and Howard Zinn, *The Cold War and the University: Toward an Intellectual History of the Postwar Years*, 171–194. New York: New Press.

Clarke, Adele E., and Joan H. Fujimura. 1992. "What Tools? Which Jobs? Why Right?" In *The Right Tools for the Job: At Work in Twentieth-Century Life Sciences*, ed. Adele E. Clarke and Johan Fujimura, 3–44. Princeton, NJ: Princeton University Press.

Clifford, James. 1978. "Hanging Up Looking Glasses at Odd Corners": Ethnobiographical Prospects." In *Studies in Biography*, ed. D. Aaron, 41–56. Cambridge, MA: Harvard University Press.

Cobb, Stanley. 1960. "A Salute from Neurologists." In *The Neuropsychology of Lashley*, ed. Frank A. Beach, Donald O. Hebb, Clifford T. Morgan and Henry W. Nissen, xvii–xx. New York: McGraw-Hill.

Cobb, Stanley. 1952. "On the Nature and Locus of Mind." *A.M.A. Archives of Neurology and Psychiatry* 67 (2): 172–177.

Cohen-Cole, Jamie. 2014. *The Open Mind: Cold War Politics and the Sciences of Human Nature*. Chicago: University of Chicago Press.

Cohen-Cole, Jamie. 2009. "The Creative American: Cold War Salons, Social Science, and the Cure for Modern Society." *Isis* 100 (2): 219–262.

Cohen-Cole, Jamie. 2007. "Instituting the Science of Mind: Intellectual Economies and Disciplinary Exchange at Harvard's Center for Cognitive Studies." *British Journal for the History of Science* 40 (4): 567–597.

Cohen-Cole, Jamie. 2005. "The Reflexivity of Cognitive Science: The Scientist as a Model of Human Nature." *History of the Human Sciences* 18 (4): 107–139.

Collins, Alan F. 2012. "An Asymmetric Relationship: The Spirit of Kenneth Craik and the Work of Warren McCulloch." *Interdisciplinary Science Reviews* 37 (3): 254–268.

Commission on Medical Education. 1932. *Final Report of the Commission on Medical Education*. New York: Office of the Director of Study.

Conway, Flo, and Jim Siegelman. 2005. *Dark Hero of the Information Age: In Search of Norbert Wiener, the Father of Cybernetics*. New York: Basic Books.

Coon, Horace. 1947. *Columbia: Colossus on the Hudson*. New York: E.P. Dutton & Co.

Cordeschi, Roberto. 2002. *The Discovery of the Artificial: Behavior, Mind and Machines Before and Beyond Cybernetics*. Dordrecht: Kluwer.

Craik, Kenneth J. W. 1943. *The Nature of Explanation*. Cambridge: Cambridge University Press.

Creager, Angela N. H. 2002. *The Life of a Virus: Tobacco Mosaic Virus as an Experimental Model, 1930–1965*. Chicago: University of Chicago Press.

Creager, Angela N. H., M. Norton Wise, and Elizabeth Lunbeck, eds. 2007. *Science Without Laws: Model Systems, Cases, Exemplary Narratives*. Durham, NC: Duke University Press.

Crevier, Daniel. 1993. *AI: The Tumultuous History of the Search for Artificial Intelligence*. Basic Books.

Crombie, Alastair C. 1994. *Styles of Scientific Thinking in the European Tradition: The History of Argument and Explanation Especially in the Mathematical and Biomedical Sciences and Arts*. London: Duckworth.

Crowther-Heyck, Hunter. 2006. "Patrons of the Revolution: Ideals and Institutions in Postwar Behavioral Science." *Isis* 97 (3): 420–446.

Crowther-Heyck, Hunter. 2005. *Herbert A. Simon: The Bounds of Reason in Modern America*. Baltimore, MD: Johns Hopkins University Press.

Crowther-Heyck, Hunter. 1999. "George A. Miller, Language, and the Computer Metaphor of Mind." *History of Psychology* 2:37–64.

Cunningham, Andrew, and Perry Williams, eds. 1992. *The Laboratory Revolution in Medicine*. Cambridge: Cambridge University Press.

Currell, Susan. 2006. "Eugenic Decline and Recovery in Self-Improvement Literature of the Thirties." In *Popular Eugenics: National Efficiency and American Mass Culture in the 1930s*, ed. Susan Currell and Christina Cogdell, 44–69. Athens, OH: Ohio University Press.

Danziger, Kurt. 1997. *Naming the Mind: How Psychology Found Its Language*. London: Sage.

Danziger, Kurt. 1990. *Constructing the Subject: Historical Origins of Psychological Research*. Cambridge: Cambridge University Press.

Danziger, Kurt. 1987. "Social Context and Investigative Practice in Early Twentieth-Century Psychology." In *Psychology in Twentieth-Century Thought and Society*, ed. Mitchell G. Ash and William R. Woodward, 13–33. Cambridge: Cambridge University Press.

Darrow, Chester W., Warren S. McCulloch, J. R. Greene, E. W. Davis, and Hugh W. Garol. 1944. "Parasympathetic Regulation of High Potential in the Electroencephalogram." *Journal of Neurophysiology* 7:217–226.

Darrow, Chester W., J. R. Greene, Warren S. McCulloch. 1944. "Activity in the Great Superficial Petrosal Nerve Influencing the Electroencephalogram." *Federation Proceedings* 3 (1): 8.

Daston, Lorraine, and Otto H. Sibum. 2003. "Scientific Personae." *Science in Context* 16 (1): 1–18.

Davis, E. W., Warren S. McCulloch, and E. Roseman. 1944. "Rapid Changes in the $O_2$ Tension of Cerebral Cortex during Induced Convulsions." *American Journal of Psychiatry* 100:825–829.

Deignan, Stella Leche, and Esther Miller. March 28, 1952. "The Support of Research in Medical and Allied Fields for the Period 1946 through 1951." *Science* 115 (2987): 321–343.

DiPalma, Anthony. 1992. "Prof. F. S. C. Northrop Dies at 98; Saw Conflict Resolution in Science." *New York Times*, July 23.

Dowie, Mark. 2001. *American Foundations: An Investigative History.* Cambridge, MA: MIT Press.

Droogleever-Fortuyn, J. 1989. "Reticular Formation and Choice of Behavior." In *The Collected Works of Warren S. McCulloch*, vol. 4, ed. Rook McCulloch, 1302–1305. Salinas, CA: Intersystems.

Dumenil, Lynn. 1995. *The Modern Temper: American Culture and Society in the 1920s.* New York: Hill and Wang.

Dupree, A. Hunter. 1972. "The Great Instauration of 1940: The Organization of Scientific Research for War." In *The Twentieth-Century Sciences: Studies in the Biography of Ideas*, ed. Gerald Holton, 443–467. New York: Norton.

Dupuy, Jean-Pierre. 2000. *The Mechanization of the Mind: On the Origins of Cognitive Science.* Trans. M. B. DeBevoise. Princeton, NJ: Princeton University Press.

Dusser de Barenne, Johannes G. 1934. "Some Aspects of the Problem of 'Corticalization' of Function and of Functional Localization in the Cerebral Cortex." In *Localization of Function in the Cerebral Cortex*, ed. S. T. Orton, J. F. Fulton, and T. K. Davis, 85–106. Proceedings of the Association for Research in Nervous and Mental Disease (Vol. XIII), Baltimore, MD: Williams and Wilkins.

Dusser de Barenne, Johannes G. 1932. "Yale University, School of Medicine, Laboratory of Neurophysiology." In *Methods and Problems of Medical Education*, no. 20, 25–26. New York: Division of Medical Education, Rockefeller Foundation.

Dusser de Barenne, Johannes G. 1924. "Experimental Researches on Sensory Localization in the Cerebral Cortex of the Monkey (Macacus)." *Proceedings of the Royal Society of London. Series B, Containing Papers of a Biological Character* 96 (676): 272–291.

Dusser de Barenne, Johannes G. 1916. "Experimental Researches on Sensory Localisations in the Cerebral Cortex." *Quarterly Journal of Experimental Physiology (Cambridge, England)* 9:355–390.

Dusser de Barenne, Johannes G. 1910. "Die Strychninwirkung auf das Zentralnervensystem. I. Die Wirkung des Strychnins auf die Reflextätigkeit der Intervertebralganglia." *Folia Neuro-Biologica Leipzig* 4:467–474.

Dusser de Barenne, Johannes G., and Warren S. McCulloch. 1939. "Physiological Delimitation of Neurones in the Central Nervous System." *American Journal of Physiology* 127:620–628.

Dusser de Barenne, Johannes G., and Warren S. McCulloch. 1938. "Sensorimotor Cortex, Nucleus Caudatus, and Thalamus Opticus." *Journal of Neurophysiology* 1:364–376.

Dusser de Barenne, Johannes G., and Warren S. McCulloch. 1938. "Functional Organization in the Sensory Cortex of the Monkey (*Macaca mulatta*)." *Journal of Neurophysiology* 1:69–85.

Dusser de Barenne, Johannes G., and Warren S. McCulloch. 1938. "The Direct Functional Interrelation of Sensory Cortex and Optic Thalamus." *Journal of Neurophysiology* 1:176–186.

Dusser de Barenne, Johannes G., and Warren S. McCulloch. 1936. "Some Effects of Local Strychninization on Action Potentials of the Cerebral Cortex of the Monkey." *Transactions of the American Neurological Association* (62nd Annual Meeting, Atlantic City, NJ, June 1–3), 171.

Dusser de Barenne, Johannes G., and Warren S. McCulloch. 1934. "An 'Extinction' Phenomenon on Stimulation of the Cerebral Cortex." *Proceedings of the Society for Experimental Biology and Medicine* 32:524–527.

Dusser de Barenne, Johannes G., Warren S. McCulloch, and Teizo Ogawa. 1938. "Functional Organization in the Face-Subdivision of the Sensory Cortex of the Monkey (*Macaca mulatta*)." *Journal of Neurophysiology* 1:436–441.

Dwyer, Ellen. 2012. "Neurological Patients as Experimental Subjects: Epilepsy Studies in the United States." In *The Neurological Patient in History*, ed. L. S. Jacyna and Stephen T. Casper, 44–60. Rochester, NY: Rochester University Press.

Dwyer, Ellen. Winter 2000. "Toward New Narratives of Twentieth-Century Medicine." *Bulletin of the History of Medicine* 74 (4): 786–793.

Eckhardt, William. 1974. "Changing Concerns in Peace Research and Education." *Bulletin of Peace Proposals* 3:280.

Edwards, Paul N. 1996. *The Closed World: Computers and the Politics of Discourse in Cold War America*. Cambridge, MA: MIT Press.

Engerman, David. Summer 2003. "Rethinking the Cold War Universities: Some Recent Histories." *Journal of Cold War Studies* 5 (3): 80–95.

Erickson, Paul. 2010. "Mathematical Models, Rational Choice, and the Search for Cold War Culture." *Isis* 101 (2): 386–392.

Erickson, Paul, Judy L. Klein, Lorraine Daston, Rebecca Lemov, Thomas Sturm, and Michael D. Gordin. 2013. *How Reason Almost Lost its Mind: The Strange Career of Cold War Rationality*. Chicago: University of Chicago Press.

Erneling, Christina E., and David Martel Johnson. 1997. *The Mind as a Scientific Object: Between Brain and Culture*. Oxford University Press.

Fass, Paula S. 1979. *The Damned and the Beautiful: American Youth in the 1920s.* Oxford University Press.

Feigenbaum, Edward A., and Julian Feldman. 1963. "Preface." In *Computers and Thought,* ed. Edward A. Feigenbaum and Julian Feldman, v–vii. New York: McGraw-Hill.

Finger, Stanley. 1994. *Origins of Neurosciences: A History of Explorations into Brain Function.* New York: Oxford University Press.

Fink, Max. November 1999. "Images in Psychiatry: Ladislas J. Meduna, M.D., 1896–1964." *American Journal of Psychiatry* 156 (11): 1807.

Forbes, Shannon. Fall 2007. "Performative Identity Formation in Frank McCourt's *Angela's Ashes: A Memoir.*" *Journal of Narrative Theory* 37 (3): 473–496.

Forman, Paul. 2012. "On the Historical Forms of Knowledge Production and Curation: Modernity Entailed Disciplinarity, Postmodernity Entails Antidisciplinarity," *Osiris* 27: 56–97.

Forman, Paul. 1987. "Behind Quantum Electronics: National Security as a Basis for Physical Research in the United States, 1940–1960." *Historical Studies in the Physical Sciences* 18 (1): 149–229.

Fosdick, Raymond B. 1952. *The Story of the Rockefeller Foundation.* New York: Harper.

Frank, Lawrence K. October 13, 1948. "Foreword." *Annals of the New York Academy of Sciences* 50 (4): 189–196.

Frank, Philipp. 1951. "Introductory Remarks, Contributions to the Analysis and Synthesis of Knowledge." *Proceedings of the American Academy of Arts and Sciences* 80:5–8.

Frank, Philipp. 1947. "The Institute for the Unity of Science: Its Background and Purpose." *Synthese* 6 (3/4—July/August): 160–167.

Frank, Robert G. 1994. "Instruments, Nerve Action, and the All-or-None Principle." *Osiris* 9:208–235.

Fremont-Smith, Frank. "Introductory Discussion." In *Cybernetics: Causal, and Feedback Mechanisms in Biological and Social Systems: Transactions of the Sixth Conference,* ed. Heinz von Foerster, 9–26. New York: Josiah Macy Jr. Foundation, 1949. In Claus Pias (ed.), *Cybernetics—Kybernetik: The Macy Conferences 1946–1953* (Zürich-Berlin: Diaphanes, 2003), 29–40.

Friedlander, Walter J. 2000. "The Rise and Fall of Bromide Therapy in Epilepsy." *Archives of Neurology* 57 (12): 1782–1785.

Fujimura, Joan. 1996. "Standardizing Practices: A Socio-History of Experimental Systems in Classical Genetic and Virological Cancer Research, ca. 1920–1978." *History and Philosophy of the Life Sciences* 18:3–54.

Fuller, Steve. 2014. "Neuroscience, Neurohistory, and the History of Science: A Tale of Two Brain Images." *Isis* 105 (1): 100–109.

Fuller, Steve. 1991. "Disciplinary Boundaries and the Rhetoric of the Social Sciences." *Poetics Today* 12 (2): 301–325.

Fulton, John F., and Ralph W. Gerard. 1940. "J. G. Dusser de Barenne 1885–1940." *Journal of Neurophysiology* 3:283–292.

Gach, John. 1980. "Culture and Complex: On the Early History of Psychoanalysis in America." In *Essays in the History of Psychiatry*, ed. Edwin R. Wallace, IV, and Lucius Pressley, 135–160. Columbia, SC: William S. Hall Psychiatric Institute.

Gaito, John. 1958. "Neuropsychological Approach to Thinking." *Psychological Reports* 4 (2): 323–332.

Galison, Peter. 2001. "The Americanization of Unity." In *Science in Culture*, ed. Peter Galison, S. R. Graubard, and Everett Mendelsohn, 45–71. New Brunswick/London: Transaction Publishers.

Galison, Peter. 1997. *Image and Logic: A Material Culture of Microphysics*. Chicago: University of Chicago Press.

Galison, Peter. 1994. "The Ontology of the Enemy: Norbert Wiener and the Cybernetic Vision." *Critical Inquiry* 21 (1): 228–266.

Gardner, Howard. 1985. *The Mind's New Science*. New York: Basic Books.

Gavrus, Delia. 2011. "Men of Dreams and Men of Action: Neurologists, Neurosurgeons, and the Performance of Professional Identity, 1925–1950." *Bulletin of the History of Medicine* 85 (1): 57–92.

Gavrus, Delia. October 2006. "Mind over matter: Sherrington, Penfield, Eccles, Walshe, and the Dualist Movement in Neuroscience." *MCIS Briefings: Comparative Program on Health and Society, Lupina Foundation, Working Papers Series 2005–2006*: 51–75.

Geertz, Clifford. 2000. "The Way We Think Now: Towards an Ethnography of Modern Thought." In Clifford Geertz, *Local Knowledge: Further Essays in Interpretative Anthropology*, 147–163. New York: Basic Books.

Geiger, Roger L. 1993. *Research and Relevant Knowledge: American Research Universities Since World War II*. New York: Oxford University Press.

Geiger, Roger L. 1992. "Science, Universities, and National Defense, 1945–1970," *Osiris* 2nd Ser. Vol 7: 26–48.

Geiger, Roger L. Jan.–Feb. 1990. "Organized Research Units—Their Role in the Development of University Research." *Journal of Higher Education* 61 (1): 1–19.

Geison, Gerald L., ed. 1987. *Physiology in the American Context, 1850–1940*. Baltimore, MD: American Physiological Society.

Geison, Gerald L. 1979. "Divided We Stand: Physiologists and Clinicians in the American Context." In *The Therapeutic Revolution: Essays in the Social History of American Medicine*, ed. Morris J. Vogel and Charles E. Rosenberg, 67–90. Philadelphia: University of Pennsylvania Press.

Gerard, Ralph W. 1970. "Warren Sturgis McCulloch: Rebel Genius." *Transactions of the American Neurological Association* 95:344–345.

Gerard, Ralph W. 1951. "Some of the Problems Concerning Digital Notions in the Central Nervous System." In *Cybernetics: Circular Causal, and Feedback Mechanisms in Biological and Social Systems: Transactions of the Seventh Conference*, ed. Heinz von Foerster, Margaret Mead, and Hans Lukas Teuber (New York: Josiah Macy Jr. Foundation), 11–57, 11. In Claus Pias (ed.), *Cybernetics—Kybernetik: The Macy Conferences 1946–1953* (Zürich-Berlin: Diaphanes, 2003), 171–202, 171.

Ghamari-Tabrizi, Sharon. 2005. *The Worlds of Herman Kahn: The Intuitive Science of Thermonuclear War*. Cambridge, MA/London: Harvard University Press.

Gieryn, Thomas. 1999. *Cultural Boundaries of Science: Credibility on the Line*. Chicago: University of Chicago Press.

Gieryn, Thomas. 1983. "Boundary Work and the Demarcation of Science from Non-Science: Strains and Interests in Professional Ideologies of Scientists." *American Sociological Review* 48:781–795.

Gingras, Yves. 2001. "Pour une biographie sociologique." *Revue d'Histoire de l'Amerique Francaise* 45 (1): 123–131.

Giri, Ananta Kumar. 1998. "Transcending Disciplinary Boundaries: Creative Experiments and the Critiques of Modernity." *Critique of Anthropology* 18 (4): 379–404.

Goertz, Christopher G. 1996. "Early American Professorships in Neurology." *Annals of Neurology* 40 (2): 258–263.

Golinski, Jan. 2016. *The Experimental Self: Humphry Davy and the Making of a Man of Science*. Chicago: University of Chicago Press.

Golinski, Jan. 2011. "Humphry Davy: The Experimental Self." *Eighteenth-Century Studies* 45 (1): 15–28.

Good, James M. M. Fall 2000. "Disciplining Social Psychology: A Case Study of Boundary Relations in the History of the Human Sciences." *Journal of the History of the Behavioral Sciences* 36 (4): 383–403.

Graham, David, and Kate Sontag. 2001. "Containing Multitudes." In *After Confession: Poetry as Autobiography*, ed. Kate Sontag and David Graham, 3–8. Saint Paul, MN: Graywolf Press.

Green, Christopher D. 1995. "All That Glitters: A Review of Psychological Research on the Aesthetics of the Golden Section." *Perception* 24:937–968.

Greenblatt, Stephen. 1980. *Renaissance Self-Fashioning: From More to Shakespeare.* Chicago: University of Chicago Press.

Grob, Gerald N. 1991. *From Asylum to Community: Mental Health Policy in Modern America.* Princeton, NJ: Princeton University Press.

Grob, Gerald N. 1990. "World War II and American Psychiatry." *Psychohistory Review* 19:41–69.

Grob, Gerald N. 1987. "The Forging of Mental Health Policy in America: World War II to the New Frontier." *Journal of the History of Medicine and Allied Sciences* 42:410–446.

Grob, Gerald N. 1985. *The Inner World of American Psychiatry, 1890–1940: Selected Correspondence.* New Brunswick, NJ: Rutgers University Press.

Grob, Gerald N. 1983. *Mental Illness and American Society: 1870–1940.* Princeton, NJ: Princeton University Press.

Gross, Charles G. 2009. *A Hole in the Head: More Tales in the History of Neuroscience.* Cambridge, MA: MIT Press.

Gunderson, Keith. 1967. "Cybernetics." In *The Encyclopedia of Philosophy*, ed. Paul Edwards, 280–284. New York/London: Macmillan.

Hacking, Ian. 1996. "The Disunities of the Sciences." In *The Disunity of Science: Boundaries, Contexts, and Power*, ed. Peter Galison and David J. Stump, 37–74. Stanford, CA: Stanford University Press.

Hacking, Ian. 1992. "'Style' for Historians and Philosophers." *Studies in History and Philosophy of Science* 23 (1): 1–20.

Hagner, Michael, and Cornelius Borck. 2001. "Mindful Practices: On the Neurosciences in the Twentieth Century." *Science in Context* 14 (4): 507–510.

Hale, Nathan G., Jr. 1995. *The Rise and Crisis of Psychoanalysis in the United States: Freud and the Americans 1917–1995.* New York: Oxford University Press.

Hambidge, Jay. 1926. *Elements of Dynamic Symmetry.* New York: Brentano.

Hankins, Thomas L. 1979. "In Defence of Biography: The Use of Biography in the History of Science." *History of Science* 17:1–16.

Hardcastle, Gary L. 2003. "Debabelizing Science: The Harvard Science of Science Discussion Group, 1940–44." In *Logical Empiricism in North America*, ed. Gary L. Hardcastle and Alan Richardson, 170–196. Minneapolis, MN: University of Minnesota Press.

Hardcastle, Gary L., and Alan W. Richardson, eds. 2003. *Logical Empiricism in North America* (Minnesota Studies in the Philosophy of Science, Volume XVIII). Minneapolis, MN: University of Minnesota Press.

Harman, Oren. 2011. "Helical Biography and the Historical Craft: The Case of Altruism and George Price." *Journal of the History of Biology* 44:671–691.

Harman, Oren. 2010. *The Price of Altruism: George Price and the Search for the Origins of Kindness.* New York: W.W. Norton.

Harrington, Anne. 1987. *Medicine, Mind and the Double Brain: A Study in 19th-Century Thought.* Princeton, NJ: Princeton University Press.

Hatfield, Gary. 1990. *The Natural, and the Normative: Theories of Spatial Representation from Kant to Helmholtz.* Cambridge, MA: MIT Press.

Hayles, N. 1999. *Katherine. How We Became Posthuman: Virtual Bodies in Cybernetics, Literature, and Informatics.* Chicago: University of Chicago Press.

Heims, Steve J. 1993. *Constructing a Social Science for Postwar America: The Cybernetics Group, 1946–1953.* Cambridge, MA: MIT Press.

Hempel, Carl G., and Paul Oppenheim. 1948. "Studies in the Logic of Explanation." *Philosophy of Science* 15 (1): 135–175.

Herman, Morris. 1970. "S. Bernard Wortis 1904–1969." *Transactions of the American Neurological Association* 95:343–344.

Herzig, Rebecca. 2004. "On Performance, Productivity, and Vocabularies of Motive in Recent Studies of Science." *Feminist Theory* 5:127–147.

Heyck, Hunter. 2015. *The Age of System: Understanding the Development of Modern Social Science.* Baltimore, MD: Johns Hopkins University Press.

Heyck, Hunter. 2014. "The Organizational Revolution and the Human Sciences." *Isis* 105 (1): 1–31.

Heyck, Hunter, and David Kaiser. 2010. "'Introduction.' Focus: New Perspectives on Science and the Cold War." *Isis* 101 (2): 362–366.

Hilgard, E. R., and C. P. Stone. 1939. "Physiological Psychology." *Annual Review of Physiology* 1:471–487.

Hill, Archibald. 1910. "A New Mathematical Treatment of Changes of Ionic Concentration in Muscle and Nerve under the Action of Electric Currents, with a Theory as to Their Mode of Excitation." *Journal of Physiology* 40:190–224.

Hirshbein, Laura. 2012. "Historical Essay: Electroconvulsive Therapy, Memory, and Self in America." *Journal of the History of the Neurosciences* 21:147–169.

Hirshbein, Laura, and Sharmalie Sarvananda. Winter 2008. "History, Power, and Electricity: American Popular Magazine Accounts of Electroconvulsive Therapy, 1940–2005." *Journal of the History of the Behavioral Sciences* 44 (1): 1–18.

Hoff, H. E. 1962. "John Fulton's Contribution to Neurophysiology." *Journal of the History of Medicine and Allied Sciences* 17:16–71.

Holton, Gerald. 1995. "On the Vienna Circle in Exile: An Eyewitness Report." In *The Foundational Debate: Complexity and Constructivity in Mathematics and Physics*, ed. Werner Depauli-Schimanovich, Eckehart Köhler, and Friedrich Stadler, 269–292. Dordrecht: Kluwer.

Hughes, John R. 1993. "Development of Clinical Psychology in the Chicago Area." *Journal of Clinical Neurophysiology* 10 (3): 298–303.

Hughlings Jackson, John 1884. "Croonian Lecture, Part 1, March 1884." *Lancet* 1: 535).

Hughlings Jackson, John. 1873. "On the Anatomical and Physiological Localization of Movements in the Brain." *The (London) Lancet*, no. 4 (April): 197–201 [Part I]; no. 5 (May): 245–48 [Part 2].

Hull, Clark L. 1943. *Principles of Behavior: An Introduction to Behavior Theory*. New York: Appleton-Century-Crofts.

Hull, Clark L. 1940. *Mathematico-Deductive Theory of Rote Learning*. New Haven, CT: Yale University Press.

Hull, Clark L. 1937. "Mind, Mechanism, and Adaptive Behavior." *Psychological Review* 44 (1): 1–32.

Isaac, Joel. 2012. *Working Knowledge: Making the Human Sciences from Parsons to Kuhn*. Cambridge, MA: Harvard University Press.

Isaac, Joel. 2011. "Introduction: The Human Sciences and Cold War America." *Journal of the History of the Behavioral Sciences* 47 (3): 225–231.

Isaac, Joel. 2007. "The Human Sciences in Cold War America." *Historical Journal (Cambridge, England)* 50 (3): 725–746.

Jacyna, L. S. 2000. *Lost Words: Narratives of Language and the Brain, 1825–1926*. Princeton, NJ: Princeton University Press.

James, William. 1890. *Principles of Psychology*, vol. I. New York: Holt.

Jeffress, Lloyd A. 1951. *Cerebral Mechanisms in Behavior: The Hixon Symposium*. New York: John Wiley.

Jewett, Andrew. 2012. *Science, Democracy, and the American University: From the Civil War to the Cold War*. Cambridge: Cambridge University Press.

Johnson, Samuel. 1796. *Rasselas, Prince of Abissinia: A Tale.* London: E. and S. Harding, Pall Mall. Chapter X. Eighteenth Century Collections Online.

Jones, Rufus M. 1933. *Haverford College: A History and Interpretation.* Macmillan.

Jones-Imhotep, Edward. 2012. "Maintaining Humans." In *Cold War Social Science: Knowledge Production, Liberal Democracy, and Human Nature,* ed. Mark Solovey and Hamilton Cravens, 175–195. New York: Palgrave Macmillan.

Josiah Macy Jr. Foundation. 1950. *Twentieth Anniversary Review of the Josiah Macy Jr. Foundation.* New York: The Josiah Macy Jr. Foundation.

Josiah Macy Jr. Foundation. 1955. *The Josiah Macy Jr. Foundation 1930–1955. A Review of Activities.* New York: The Josiah Macy Jr. Foundation.

Kaiser, David. 2010. "Elephant on the Charles: Postwar Growing Pains". In *Becoming MIT: Moments of Decision,* ed. David Kaiser, 103–121. Cambridge, MA: MIT Press.

Kannerstein, Gregory, ed. 1983. *The Spirit and the Intellect: Haverford College 1833–1983.* Haverford, PA: Haverford College.

Kay, Lily E. 2001. "From Logical Neurons to Poetic Embodiments of Mind: Warren S. McCulloch's Project in Neuroscience." *Science in Context* 14 (4): 591–614.

Kay, Lily E. 2000. *Who Wrote the Book of Life? A History of the Genetic Code.* Stanford, CA: Stanford University Press.

Kay, Lily E. 1993. *The Molecular Vision of Life: Caltech, the Rockefeller Foundation, and the Rise of the New Biology.* New York: Oxford University Press.

Keller, Evelyn Fox. 2002. *Making Sense of Life: Explaining Biological Development with Models, Metaphors, and Machines.* Cambridge, MA: Harvard University Press.

Keller, Evelyn Fox. 1995. *Refiguring Life: Metaphors of Twentieth-Century Biology.* New York: Columbia University Press.

Kelley, Brooks Mather. 1974. *Yale: A History.* New Haven: Yale University Press.

Keränen, Lisa. 2010. *Scientific Characters: Rhetoric, Politics, and Trust in Breast Cancer Research.* Tuscaloosa, AL: University of Alabama Press.

Kevles, Daniel J. 1977. "The National Science Foundation and the Debate over Postwar Research Policy, 1942–1945." *Isis* 68:5–26.

Kilmer, William, and Warren S. McCulloch. 1989. "The Reticular Formation Command and Control System." In *Information Processing in the Nervous System* (Proceedings of the Symposium on Information Processing in the Nervous System, State University of New York at Buffalo, 21–24 October 1969), ed. K. N. Leibovic, 99–102. Springer-Verlag, 1969. In *The Collected Works of Warren S. McCulloch, Volume 4,* ed. Rook McCulloch, 1322–1332. Salinas, CA: Intersystems.

Kilmer, William L., Warren S. McCulloch, and J. Blum, with E. Craighill and D. Peterson. 1968. "Towards a Theory of the Reticular Formation." In *The Mind: Biological Approaches to its Functions*, ed. William C. Corning and Martin Balaban, 177–232. New York: Intersciences.

Klee, Robert. 1993. "The Phrenetic Calculus: A Logician's View of Disordered Logical Thinking in Schizophrenia." *Behavior and Philosophy* 20/21 (2/1): 49–61.

Klein, Julie Thompson. 2013. "The Transdisciplinary Moment(um)." *Integral Review* 9 (2): 189–199.

Klein, Julie Thompson, ed. 2001. *Transdisciplinarity: Joint Problem Solving Among Science, Technology, and Society.* Birkhäuser.

Klein, Julie Thompson. 1996. *Crossing Boundaries: Knowledge, Disciplinarities, and Interdisciplinarities.* Charlottesville/London: University Press of Virginia.

Klein, Julie Thompson. 1993. "Blurring, Cracking, and Crossing: Permeation and the Fracturing of Discipline." In *Knowledges: Historical and Critical Studies in Disciplinarity*, ed. Ellen Messer-Davidow, David R. Shumnay, and David J. Sylvan, 185–211. Charlottesville, London: University of Virginia Press.

Klein, Julie Thompson. 1990. *Interdisciplinarity: History, Theory, and Practice.* Detroit: Wayne State University Press.

Kline, Ronald R. 2015. *The Cybernetics Moment: Or Why We Call Our Age the Information Age.* Baltimore, MD: Johns Hopkins University Press.

Kline, Ronald R. 2011. "Cybernetics, Automata Studies, and the Dartmouth Conference on Artificial Intelligence." *IEEE Annals of the History of Computing* (Oct/Dec): 5–16.

Kline, Ronald. June 2009. "Where Are the Cyborgs in Cybernetics?" *Social Studies of Science* 33:331–362.

Kneebone, G. T. 1963. *Mathematical Logic and the Foundations of Mathematics.* London: D. van Nostrand.

Kneeland, Timothy W., and Carol A. B. Warren. 2002. *Pushbutton Psychiatry: A History of Electroshock in America.* Westport, CT: Praeger.

Kohler, Robert E. 1994. *Lords of the Fly: Drosophila Genetics and the Experimental Life.* Chicago: University of Chicago Press.

Kohler, Robert E. 1991. *Partners in Science: Foundations and Natural Scientists, 1900–1945.* Chicago: University of Chicago Press.

Kohler, Robert E. 1991. "Systems of Production: Drosophila, Neurospora, and Biochemical Genetics." *Historical Studies in the Physical and Biological Sciences* 22:87–130.

Kohler, Robert E. 1982. *From Medical Chemistry to Biochemistry: The Making of a Biomedical Discipline*. Cambridge: Cambridge University Press.

Kroker, Kenton. 2007. *The Sleep of Others and the Transformation of Sleep Research*. Toronto: University of Toronto Press.

Kubie, Lawrence S. 1930. "A Theoretical Application to Some Neurological Problems of the Properties of Excitation Waves Which Move in Closed Circuits." *Brain* 53:166–177.

Kuklick, Bruce. 2004. "Philosophy at Yale in the Century after Darwin." *History of Philosophy Quarterly* 21 (3): 313–336.

Kuklick, Bruce. 2001. *A History of Philosophy in America, 1720–2000*. Oxford: Clarendon Press.

Kuklick, Bruce. 1977. *The Rise of American Philosophy: Cambridge, Massachusetts 1860–1930*. New Haven, CT: Yale University Press.

Kusch, Martin. 2010. "Hacking's Historical Epistemology: A Critique of Styles of Reasoning." *Studies in History and Philosophy of Science* 41:158–173.

Lamb, Susan D. 2014. *Pathologist of the Mind: Adolf Meyer and the Origins of American Psychiatry*. Baltimore, MD: Johns Hopkins University Press.

Lashley, Karl. 1952. "Functional Interpretation of Anatomic Patterns in *Patterns of Organization in the Central Nervous System*." *Research Publications—Association for Research in Nervous and Mental Disease* XXX:529–547.

Lashley, Karl. 1926. "Studies of Cerebral Function in Learning VII: The Relation between Cerebral Mass, Learning, and Retention." *Journal of Comparative Neurology* 41:1–58.

Laslett, P. 1950. *The Physical Basis of Mind: A Symposium*. New York: Macmillan, Oxford: Basil Blackwell.

Laszlo, Alejandra. 1987. "Physiology of the Future: Institutional Styles at Columbia and Harvard." In *Physiology in the American Context, 1850–1940*, ed. Gerald L. Geison, 67–96. Baltimore, MD: American Physiological Society.

Latour, Bruno, and Steve Woolgar. [1979] 1986. *Laboratory Life: The Construction of Scientific Facts*. 2nd ed. Princeton, NJ: Princeton University Press.

Lawrence, Christopher, and Mayer, Anna-K. 2000. "Regenerating England: An Introduction." In *Regenerating England: Science, Medicine, and Culture in Interwar Britain*, ed. Christopher Lawrence and Anna-K. Mayer, 1–23. Amsterdam: Rodopi.

Lazenby, Jill A. 2003. "Climates of Collaboration. Interdisciplinary Science and Social Identity." PhD Dissertation, University of Toronto.

Leahy, T. 1992. "The Mythical Revolutions of American Psychology." *American Psychologist* 47:308–318.

Leiber, Justin. 1991. *An Invitation to Cognitive Science*. Oxford: Basil Blackwell.

Lemov, Rebecca. 2005. *World as Laboratory: Experiments with Mice, Mazes, and Men*. New York: Farrar, Straus, and Giroux.

Lenoir, Timothy, ed. 1997. *Instituting Science: The Cultural Production of Scientific Disciplines*. Stanford: Stanford University Press.

Leonelli, Sabina. 2007. "What's in a Model?" In *Modeling Biology: Structures, Behaviors, Evolution*, ed. Manfred D. Laubichler and Gerd M. Müller, 13–36. Cambridge, MA: MIT Press.

Leslie, Stuart W. 1994. "Science and Politics in Cold War America." In *The Politics of Western Science, 1964–1990*, ed. Margaret C. Jacob, 199–233. Humanity Books.

Leslie, Stuart W. 1993. *The Cold War and American Science: The Military-Industrial-Academic Complex at MIT and Stanford*. New York: Columbia University Press.

Leslie, Stuart W. 1990. "Profit and Loss: The Military and MIT in the Postwar Era." *Historical Studies in the Physical and Biological Sciences* 21 (1): 59–85.

Lettvin, Jerome Y. 1998. "Jerome Lettvin." In *The History of Neuroscience in Autobiography*, vol. II, ed. L. R. Squire, 222–243. San Diego: Academic Press.

Lettvin, Jerome Y. 1998. *Talking Nets: An Oral History of Neural Networks*, ed. James A. Anderson and Edward Rosenfeld, 1–21. Cambridge, MA: MIT Press. [Interview with J. A. Anderson and E. Rosenfeld]

Lettvin, Jerome Y. 1989. "Strychnine Neuronography." In *The Collected Works of Warren S. McCulloch*, vol. I, ed. Rook McCulloch, 50–58. Salinas, CA: Intersystems Publications.

Lettvin, Jerome Y. 1989. "Warren and Walter." In *The Collected Works of Warren S. McCulloch*, vol. II, ed. Rook McCulloch, 515–529. Salinas, CA: Intersystems Publications.

Lettvin, Jerome Y. 1989. "Warren as a Teacher." In *The Collected Works of Warren S. McCulloch*, vol. I, ed. Rook McCulloch, 332–338. Salinas, CA: Intersystems.

Levins, Richard. 1966. "The Strategy of Model Building in Population Biology." *American Scientist* 54:421–431.

Lewontin, Richard C. 1997. "The Cold War and the Transformation of the Academy." In Noam Chomsky, R. C. Lewontin, Ira Katznelson, Laura Nader, Richard Ohmann, David Montgomery, Immanuel Wallerstein, Ray Siever, and Howard Zinn, eds., *The Cold War and the University: Toward and Intellectual History of the Postwar Years*, 1–34. New York: New Press.

Lindsey, Ben B., and Wainright Evans. 1925. *The Revolt of Modern Youth*. New York: Boni & Liveright.

Lipowski, Zbigniew J. 1986. "Psychosomatic Medicine: Past and Present Part I: Historical Background." *Canadian Journal of Psychiatry* 31:1–7.

Lillie, Ralph S. 1931. "Types of Physical Determination and the Activities of Organisms." *Journal of Philosophy* 28 (21): 561–573.

Litt, Willard D., ed. 1921. *History of the Class of Nineteen Hundred and Twenty-One*. New Haven, CT: Tuttle, Morehouse, and Taylor.

Littlefield, Melissa M. 2010. "Matter for Thought: The Psychon in Neurology, Psychology, and American Culture, 1927–1943." In *Neurology and Modernity*, ed. Laura Salisbury and Andrew Shail, 267–286. New York: Palgrave.

Littlefield, Melissa, and Jenell M. Johnson, eds. 2012. *The Neuroscientific Turn: Transdisciplinarity in the Age of the Brain*. University of Michigan Press.

Lorente de Nó, Rafael. 1935. "Facilitation of Motoneurones." *American Journal of Physiology* 113:505–522.

Lorente de Nó, Rafael. 1933. "Vestibulo-Ocular Reflex Arc." *Archives of Neurology and Psychiatry* 30:245–291.

Lorente de Nó, Rafael. 1933. "Studies on the Structure of the Cerebral Cortex I. The Area Entorhinalis." *Journal für Psychologie und Neurologie* 45 (6): 381–438.

Lotka, Alfred. 1925. *Elements of Physical Biology*. Baltimore MD: Williams and Wilkins.

Löwy, Ilana. 1994. "Experimental Systems and Clinical Practices: Tumor Immunology and Cancer Immunotherapy, 1895–1980." *Journal of the History of Biology* 27:403–435.

Löwy, Ilana. 1992. "The Strength of Loose Concepts—Boundary Objects, Federative Experimental Strategies, and Disciplinary Growth: The Case of Immunology." *History of Science* 30:371–396.

Lunbeck, Elizabeth. 1994. *The Psychiatric Persuasion: Knowledge, Gender, and Power in Modern America*. Princeton, NJ: Princeton University Press.

Magnus, Otto. 2002. *Rudolf Magnus, Physiologist and Pharmacologist: A Biography*. Amsterdam: Royal Netherlands Academy of Arts and Sciences.

Magnus, Rudolf. 1950. *Goethe as a Scientist*. Trans. H. Norden. New York: Henry Schuman.

Magnus, Rudolf. 1930. "The Physiological a Priori." *Lane Lectures on Experimental Pharmacology and Medicine, Stanford University Publications, University Series, Medical Sciences*, Volume II, No. 3, 97–103. Stanford, CA: Stanford University Press.

Magnus, Rudolf. 1924. *Körperstellung.* Berlin: Julius Springer.

Magoun, Horace W. 2002. *American Neuroscience in the Twentieth Century: Confluence of the Neural, Behavioral, and Communicative Streams.* Swets and Zeitlinger.

Magoun, Horace W. 1975. "The Role of Research Institutes in the Advancement of Neuroscience: Ranson's Institute of Neurology." In *The Neurosciences: Paths of Discovery,* ed. Frederic G. Worden, Judith P. Swazey, and George Adelman, 515–527. Cambridge, MA: MIT Press.

Magoun, Horace W. 1952. "An Ascending Reticular Activating System in the Brain Stem." *Archives of Neurology and Psychiatry* 67 (2): 145–154.

Mahon, Basil. 2003. *The Man Who Changed Everything: The Life of James Clerk Maxwell.* New York: Wiley.

Malzberg, Benjamin. 1938. "Outcome of Insulin Treatment of One Thousand Patients with Dementia Praecox." *Psychiatric Quarterly* 12:528–553.

Marchand, Roland. 1985. *Advertising the American Dream: Making Way for Modernity, 1920–1940.* Berkeley, CA: University of California Press.

Marks, Harry M. 1997. *The Progress of Experiment: Science and Therapeutic Reform in the United States, 1900–1990.* Cambridge University Press.

Marshall, Louise H. 1987. "Instruments, Techniques, and Social Units in American Neurophysiology, 1870–1950." In *Physiology in the American Context: 1850–1940,* ed. Gerald L. Geison, 351–369. Bethesda, MD: American Physiological Society.

Marshall, Louise H., and Horace W. Magoun, eds. 1998. *Discoveries in the Human Brain: Neuroscience Prehistory, Brain structure, and Function.* Totowa, NJ: Humana Press.

May, Mark A. 1971. "A Retrospective View of the Institute of Human Relations at Yale." *Behavior Science Notes* 6:141–172.

Mayer, Anna-K. 2000. "'A Combative Sense of Duty': Englishness and the Scientists." In *Regenerating England: Science, Medicine, and Culture in Interwar Britain,* ed. Christopher Lawrence and Anna-K. Mayer. 67–106. Amsterdam, Atlanta, GA: Rodopi.

Mazzotti, Massimo. 2012. *The World of Maria Gaetana Agnesi, Mathematician of God.* Baltimore: Johns Hopkins University Press.

McCarthy, John. 1987. "Review of Brian P. Boomfield (ed.), *The Question of Artificial Intelligence: Philosophical and Sociological Perspectives.*" London/New York/Sydney: Croom Helm (http://www-formal.stanford.edu/jmc/reviews/bloomfield/bloomfield.html, accessed 11 July 2013).

McCarthy, John, Marvin L. Minsky, Nathaniel Rochester, and Claude E. Shannon. 1955. "A Proposal for the Dartmouth Summer Research Project on Artificial

Intelligence, August 31." (http://www-formal.stanford.edu/jmc/history/dartmouth/dartmouth.html; accessed 11 July 2013).

McCorduck, Pamela. 2004. *Machines Who Think: A Personal Inquiry into the History and Prospects of Artificial Intelligence*. Natick, MA: AK Peters.

McCulloch, Rook. 1989. "Foreword." In *The Collected Works of Warren S. McCulloch*, vol. 1, ed. Rook McCulloch, 1–6. Salinas, CA: Intersystems.

McCulloch, Warren S. 1989. "The Brain as a Computing Machine." In *The Collected Works of Warren S. McCulloch*, vol. 2, ed. Rook McCulloch, 582–596. Salinas, CA: Intersystems.

McCulloch, Warren S. 1989. "Recollections of the Many Sources of Cybernetics." In *The Collected Works of Warren S. McCulloch*, vol. 1, ed. Rook McCulloch, 21–49. Salinas, CA: Intersystems. Reprinted from *ASC Forum*, Volume VI, #2, Summer 1974, 5–16.

McCulloch, Warren S. 1988. *Embodiments of Mind*. 2nd edition. Cambridge, MA: MIT Press.

McCulloch, Warren S. 1969. "Interview with Percy Saltzman," *The Day It Is*, Canadian Broadcasting Corporation.

McCulloch, Warren S. 1969. "Of I and It." *Perspectives in Biology and Medicine* 12 (4): 547–560.

McCulloch, Warren S. 1967. "Lekton, Being a Belated Introduction to the Thesis of Eilhard von Domarus." In *Communication: Theory and Research. Proceedings of the First Symposium*, ed. L. Thayer, 348–350. Springfield, IL: Charles C. Thomas.

McCulloch, Warren S. 1967. "Commentary." In *Communication: Theory and Research. Proceedings of the First Symposium*, ed. L. Thayer, 412–428. Springfield, IL: Charles C. Thomas.

McCulloch, Warren S. 1965. "What's in the Brain That Ink May Character?" in Warren S. McCulloch, *Embodiments of Mind* (Cambridge, MA: MIT Press), 387–397. (Originally presented at the International Congress for Logic, Methodology, and Philosophy of Science, Jerusalem, Israel, August 28, 1964.)

McCulloch, Warren S. 1964. "Reliable Systems Using Unreliable Units." In *Disorders of Communication* (Volume XLII, Research Publications of the Association for Research in Nervous and Mental Disease), ed. David McKenzie Rioch and Edwin A. Weinstein, 19–28. Baltimore, MD: Williams and Wilkins. In *The Collected Works of Warren S. McCulloch*, vol. 4, ed. Rook McCulloch, 1292–1301. Salinas, CA: Intersystems, 1989.

McCulloch, Warren S. 1961. "Abracadabra." In *Mental Patients in Transition*, ed. M. Greenblatt, D. J. Levinson, and G. J. Klerman. Springfield, IL: Charles C. Thomas. In

*The Collected Works of Warren S. McCulloch,* vol. 4, ed. Rook McCulloch, 1244–1255. Salinas, CA: Intersystems, 1989.

McCulloch, Warren S. 1961. "Where Is Fancy Bred?" In *Lectures on Experimental Psychiatry,* ed. Henry W. Brosin (Pittsburgh, PA: University of Pittsburgh Press), 311–324. In Warren S. McCulloch, *Embodiments of Mind,* 216–229. Cambridge, MA: MIT Press, 1988.

McCulloch, Warren S. 1961. "What Is a Number, That a Man May Know It, and a Man, That He May Know a Number?" *Embodiments of Mind,* 1–18. Cambridge, MA: MIT Press, 1988. (Originally published as The Ninth Annual Alfred Korzybski Memorial Lecture, *General Semantics Bulletin Nos. 26 and 27,* Lakeville, CT: Institute of General Semantics, 1961, 7–18.)

McCulloch, Warren S. 1960. "The Reliability of Biological Systems." In *Self-Organizing Systems,* ed. M. C. Yovitz and S. Cameron, 264–280. Pergamon Press. In *The Collected Works of Warren S. McCulloch,* vol. 4, ed. Rook McCulloch, 1193–1210. Salinas, CA: Intersystems, 1989.

McCulloch, Warren S. 1959. *The Natural Fit.* Chicago: Chicago Literary Club.

McCulloch, Warren S. 1958. "Agathe Tyche of Nervous Nets: The Lucky Reckoners," in *Mechanisation of Thought Processes: Proceedings of a Symposium Held at the National Physical Laboratory,* November 24–27, No. 10, Vol. II (London: Her Majesty's Stationery Office, 1959), 611–634. In *Embodiments of Mind,* 203–215. Cambridge, MA: MIT Press, 1988.

McCulloch, Warren S. 1955. "Introductory Remarks." In *Cybernetics: Circular Causal and Feedback Mechanisms in Biological and Social Systems: Transactions of the Tenth Conference,* ed. Heinz von Foerster, Margaret Mead, and Hans Lukas Teuber, 15–18. New York: Josiah Macy, Jr. Foundation. In *Cybernetics—Kybernetik: The Macy Conferences 1946–1953,* ed. Claus Pias, 687–688. Zürich-Berlin: Diaphanes, 2003.

McCulloch, Warren S. 1955. "Summary of the Points of Agreement Reached in the Previous Nine Conferences on Cybernetics." In *Cybernetics: Circular Causal and Feedback Mechanisms in Biological and Social Systems: Transactions of the Tenth Conference,* ed. Heinz von Foerster, Margaret Mead, and Hans Lukas Teuber. New York: Josiah Macy, Jr. Foundation, 69–80. In *Cybernetics—Kybernetik: The Macy Conferences 1946–1953,* ed. Claus Pias, 719–725. Zürich-Berlin: Diaphanes, 2003.

McCulloch, Warren S. 1955. "*Mysterium iniquiatis* of Sinful Man Aspiring into the Place of God." *The Scientific Monthly* 80, no.1: 35–39. In Warren S. McCulloch, *Embodiments of Mind,* 157–164. Cambridge, MA: MIT Press, 1988.

McCulloch, Warren S. 1952. "The Past of a Delusion." Paper Presented to the Chicago Literary Club, January, in *The Collected Works of Warren S. McCulloch, Volume 2,* ed. Rook McCulloch, 761–791. Salinas, CA: Intersystems, 1989.

McCulloch, Warren S. 1951. "Why the Mind Is in the Head." In *Cerebral Mechanisms in Behavior: The Hixon Symposium*, edited by Lloyd A. Jeffress, 42–57. New York: John Wiley.

McCulloch, Warren S. 1949. "Physiological Processes Underlying Psychoneuroses." *Proceedings of the Royal Society, Medical Supplement* XLII (Anglo-American Symposium on Psychosurgery, Neurophysiology, and Physical Treatments in Psychiatry): 72–84.

McCulloch, Warren S. 1948. "A Recapitulation of the Theory, with Forecast of Several Extensions." *Annals of the New York Academy of Sciences* 50 (4): 259–288.

McCulloch, Warren S. 1948. "Through the Den of the Metaphysician." Lecture delivered to the Philosophical Club of the University of Virginia, 23 March 1948, English version of "Dans l'antre du métaphysician," *Thales*, Volume 7 (Paris: Presses Universitaires de France, 1951), 37–49. In Warren S. McCulloch, *Embodiments of Mind*, 142–156. Cambridge, MA: MIT Press, 1988.

McCulloch, Warren S. 1945. "Modes of Functional Organization in the Cerebral Cortex." *Federation Proceedings*, 6, no. 2 (June 1945): 448–552; reprinted in *The Collected Works of Warren S. McCulloch*, vol. 3, ed. Rook McCulloch, 500–510. Salinas, CA: Intersystems, 1989.

McCulloch, Warren S. 1944. "The Functional Organization of the Cerebral Cortex." *Physiological Reviews* 24:390–407.

McCulloch, Warren S. 1940. "Johannes Gregorius Dusser de Barenne (1885–1940)." *Yale Journal of Biology and Medicine* 12:743–746.

McCulloch, Warren S., Michael A. Arbib, and Jack Cowan. 1962. "Neurological Models and Integrative Processes," in *Self-Organizing Systems*, ed. M. C. Yovitz, G. T. Jacobi, and D. G. Goldstein (Washington, DC: Spartan Books), 49–59. In *The Collected Works of Warren S. McCulloch*, vol. 4, ed. Rook McCulloch, 1260–1270. Salinas, CA: Intersystems, 1989.

McCulloch, Warren S., H. B. Carlson, and Franz G. Alexander. December 1949. "Zest and Carbohydrate Metabolism." *Research Publications—Association for Research in Nervous and Mental Disease* XXIX:406–411.

McCulloch, Warren S., and Johannes G. Dusser de Barenne. 1936. "Action Potentials of the Cerebral Cortex and Spinal Cord before and after Cortical Stimulation." *American Journal of Physiology* 118:99.

McCulloch, Warren S., and Johannes G. Dusser de Barenne. 1935. "Extinction: Local Stimulatory Inactivation within the Motor Cortex." *American Journal of Physiology* 113:97–98.

McCulloch, Warren S., C. Graf, and Horace W. Magoun. 1946. "A Cortico-Bulbo-Reticular Pathway from Area 4-s." *Journal of Neurophysiology* 9:127–132.

McCulloch, Warren S., and William L. Kilmer. 1965. "The Command and Control System of the Vertebrates." In *Proceedings of the International Federation for Information Processing, Congress, New York (May 24–29, 1965)*, Volume 2, ed. Wayne A. Kalenich, 636. Washington, DC: Spartan Books, 1966. In *The Collected Works of Warren S. McCulloch*, vol. 4, ed. Rook McCulloch, Salinas, CA: Intersystems, 1989.

McCulloch, Warren S., and W. L. Kilmer. 1964. "Toward a Theory of Reticular Formation." *5th National Symposium on Human Factors in Electronics*, 7–25.

McCulloch, Warren S., and Walter Pitts. 1943. "A Logical Calculus of the Ideas Immanent in Nervous Activity." *Bulletin of Mathematical Biophysics* 5:115–133.

McDonald, Marvin J. 1996. "Mind and Brain, Science and Religion: Belief and Neuroscience." In Donald N. Mackay and Roger W. Sperry, *Facets of Faith and Science, Volume III: The Role of Beliefs in the Natural Sciences*, ed. Jitse M. van der Meer, 199–226. Lanham, MD: University Press of America.

McWhinnie, Harold J. 1985. "A Biological Basis for the Golden Section in Art and Design." *Leonardo* 19 (3): 241–245.

Mead, Margaret. 1950. "Possible Mechanisms of Recall and Recognition." In *Cybernetics: Circular Causal, and Feedback Mechanisms in Biological and Social Systems: Transactions of the Sixth Conference*, ed. Heinz von Foerster, 146–202. New York: Josiah Macy Jr. Foundation. In Claus Pias (ed.), *Cybernetics—Kybernetik: The Macy Conferences 1946–1953* (Zürich-Berlin: Diaphanes, 2003), 122–162.

Medina, Eden. 2011. *Cybernetic Revolutionaries: Technology and Politics in Allende's Chile*. Cambridge, MA: MIT Press.

Meduna, Ladislas J. 1950. *Oneirophrenia. The Confusional State*. Urbana, IL: University of Illinois Press.

Meduna, Ladislas J. 1937. *Die Konvulsiontherapie der Schizophrenia*. Halle: Carl Marhold.

Meduna, Ladislas J., and Warren S. McCulloch. January 1945. "The Modern Concept of Schizophrenia." *The Medical Clinics of North America, Chicago Number*, 147–164. New York: W.B. Saunders.

Messer-Davidow, Ellen, David R. Shumnay, and David J. Sylvan. 1993. "Disciplinary Ways of Knowing." In *Knowledges: Historical and Critical Studies in Disciplinarity*, ed. Ellen Messer-Davidow, David R. Shumnay, and David J. Sylvan, 1–21. Charlottesville: University of Virginia Press.

Meyer, Adolf. 1957. *Psychobiology: A Science of Man*. Compiled and edited by Eunice E. Winters and Anna Mae Bowers. Springfield, IL: Charles C. Thomas.

Meyer, Adolf. 1941. "H. Douglas Singer, M.D. 1875–1940." *Archives of Neurology and Psychiatry* 45 (1): 162–164.

Meyer, Adolf. 1910. "The Dynamic Interpretation of Dementia Praecox." *American Journal of Psychology* 21:385–403.

Micale, Mark S. 2003. "The Psychiatric Body." In *Companion to Medicine in the Twentieth Century*, ed. Roger Cooter and John Pickstone, 323–346. London: Routledge.

Miller, George A. 1956. "The Magical Number Seven, Plus or Minus Two: Some Limits on Our Capacity for Processing Information." *Psychological Review* 63 (2): 81–97.

Miller, George A. October 1953. "Review of *Cybernetics: Circular Causal and Feedback Mechanisms in Biological and Social Systems, Transactions of the Eighth Conference by Heinz von Foerster, Margaret Mead, and Hans Lukas Teuber.*" *American Journal of Psychology* 66 (4): 661–663.

Miller, George A., Eugene Galanter, and Karl H. Pribham. 1960. *Plans and the Structure of Behavior*. New York: Holt.

Mindell, David. 2002. *Between Human and Machine: Feedback, Control, and Computing Before Cybernetics*. Baltimore, MD: Johns Hopkins University Press.

Mindell, David, Jérôme Ségal, and Slava Gerovitch. 2003. "From Communications Engineering to Communications Science: Cybernetics and Information Theory in the United States, France, and the Soviet Union." In *Science and Ideology: A Comparative History*, ed. Mark Walker, 66–95. London: Routledge.

Minsky, Marvin. 2011. "Warren McCulloch: For Whom the World Was a Stage." Interview with Christopher Sykes, 29–31 January, Web of Stories, (http://www.webofstories.com/play/marvin.minsky/22, accessed 10 July 2013).

Mirowski, Philip. 2002. *Machine Dreams: Economics Becomes a Cyborg Science*. Cambridge University Press.

Morawski, Jill G. 1986. "Organizing Knowledge and Behavior at Yale's Institute of Human Relations." *Isis* 77:219–242.

Morawski, Jill, ed. 1988. *The Rise of Experimentation in American Psychology*. New Haven/London: Yale University Press.

Morgan, Mary S. 2012. *The World in a Model: How Economists Work and Think*. Cambridge University Press.

Morgan, Mary S., and Till Grüne-Yanoff. 2011. "Modelling Practices in the Social and Human Sciences: An Interdisciplinary Exchange." *Perspectives on Science* 21 (2): 143–156.

Morgan, Mary S., and Margaret Morrison, eds. 1999. *Models as Mediators: Perspectives on Natural and Social Science*. Cambridge: Cambridge University Press.

Morrison, Margaret, and Mary S. Morgan. 1999. "Introduction." In *Models as Mediators: Perspectives on Natural and Social Science*, ed. Mary S. Morgan and Margaret Morrison, 1–9. Cambridge: Cambridge University Press.

Morrison, Margaret, and Mary S. Morgan. 1999. "Models as Mediating Instruments." In *Models as Mediators: Perspectives on Natural and Social Science*, ed. Mary S. Morgan and Margaret Morrison, 10–37. Cambridge: Cambridge University Press.

Moruzzi, G., and Horace W. Magoun. 1949. "Brain Stem Reticular Formation and Activation of the EEG." *Electroencephalography and Clinical Neurophysiology* 1: 455–73.

Mukerji, Chandra. 1989. *A Fragile Power: Scientists and the State*. Princeton, NJ: Princeton University Press.

Murphy, Paul V. 2012. *The New Era: American Thought and Culture in the 1920s*. Lanham, MD: Rowman & Littlefield.

Nagel, Ernst. 1953. "Teleological Explanation and Teleological Systems." In *Readings in the Philosophy of Science*, ed. Herbert Feigl and M. Brodbeck, 537–558. New York: Appleton-Century-Crofts.

Nash, Roderick. 1971. *The Nervous Generation: American Thought, 1917–1930*. Chicago: Rand McNally.

Newcomb, John Tinterman. January 2003. "The Footprint of the Twentieth Century: American Skyscrapers and Modernist Poems." *Modernism/Modernity* 10 (1): 97–125.

Newell, Alan, J. C. Shaw, and Herbert A. Simon. 1958. "Elements of a Theory of Human Problem Solving." *Psychological Review* 65:151–166.

Noll, Richard. 2011. *American Madness: The Rise and Fall of Dementia Praecox*. Cambridge, MA: Harvard University Press.

Northrop, Filmer S. C. 1989. "On W. S. McCulloch." In *The Collected Works of Warren S. McCulloch*, vol. 1, ed. Rook McCulloch, 363–369. Salinas, CA: Intersystems.

Northrop, Filmer S. C. 1947. *Logic in the Sciences and Humanities*. New York: Macmillan.

Northrop, Filmer S. C. 1940. "The Method and Theories of Physical Science and Their Bearing upon Biological Organization." *Growth (Supplement, Second Symposium on Development and Growth)* 4: 127–154.

Northrop, Filmer S. C. 1938. "The History of Modern Physics in Its Bearing on Biology and Medicine." *Yale Journal of Biology and Medicine* 10:209–232.

Northrop, Filmer S. C. 1931. *Science and First Principles*. New York: Macmillan.

Nowotny, Helga. May 2006. "The Potential of Transdisciplinarity." First published in *Interdisciplines*, http://www.inter-disciplines.org. Accessed 15 August 2014.

Nye, Mary Jo. 2011. *Michael Polanyi and His Generation: Origins of the Social Construction of Science*. Chicago: University of Chicago Press.

Nye, Mary Jo. 2006. "Scientific Biography: History of Science by Another Means?" *Isis* 97:322–329.

Nye, Mary Jo. 2004. *Blackett: Physics, War, and Politics in the Twentieth Century*. Cambridge, MA: Harvard University Press.

Opdycke, Sandra. 1999. *No One Was Turned Away: The Role of Public Hospitals in New York City Since 1900*. New York: Oxford University Press.

Oppenheim, Paul, and Hilary Putnam. 1958. "The Unity of Science as a Working Hypothesis." In *Concepts, Theories, and the Mind-Body Problem. Minnesota Studies in the Philosophy of Science Vol. II*, ed. Herbert Feigl, M. Scriven, and G. Maxwell, 3–36. Minneapolis: University of Minnesota Press.

Oppenheimer, Paul. Fall 1982. "The Origin of the Sonnet." *Comparative Literature* 34 (4): 289–304.

Otis, Laura. 2007. *Müller's Lab*. Oxford: Oxford University Press.

Papert, Seymour. 1988. "Introduction." In Warren S. McCulloch, *Embodiments of Mind*, xxi–xxviii. Cambridge, MA: MIT Press.

Parsons, Frederick W. 1927. "The Problem of Mental Hygiene in New York State." *Psychiatric Quarterly* 1 (3): 271–275.

Peirce, Charles S. 1965. "Collected Papers 5.145," *Volume V: Pragmatism and Pragmaticism*. In *Collected Papers of Charles Sanders Peirce*, ed. Charles Hartshorne and Paul Weiss. Cambridge, MA: Belknap Press/Harvard University Press.

Piccinini, Gualtiero. 2004. "The First Computational Theory of Mind and Brain: A Close Look at McCulloch and Pitts's 'Logical Calculus of Ideas Immanent in Nervous Activity.'" *Synthese* 141 (2): 175–215.

Pickering, Andrew. 2013. "Ontology and Antidisciplinarity." In *Interdisciplinarity: Reconfigurations of the Social and Natural Sciences*, ed. Andrew Barry and Georgina Born, 209–225. London: Routledge.

Pickering, Andrew. 2010. *The Cybernetic Brain: Sketches of Another Future*. Chicago: University of Chicago Press.

Pickering, Andrew. 1995. "Cyborg History and the World War II Regime." *Perspectives on Science* 3:1–48.

Pike, F. H., C. A. Eslberg, W. S. McCulloch, and M. N. Chappell. 1930. "The Problem of Localization in Experimentally Induced Convulsions." *Archives of Neurology and Psychiatry* 23 (5): 847–868.

Pike, F. H., C. A. Elsberg, W. S. McCulloch, and A. Rizzolo. 1929. "Some Observations on Experimentally Produced Convulsions: I. The Localization of the Motor Mechanism for Which the Typical Clonic Movements of Epilepsy Arise." *American Journal of Psychiatry* 9:259–283.

Pitts, Walter, and Warren S. McCulloch. 1947. "How We Know Universals: The Perception of Auditory and Visual Forms." *Bulletin of Mathematical Biophysics* 9 (3): 127–147.

Plant, Rebecca Jo. 2005. "William Menninger and American Psychoanalysis, 1946-48." *History of Psychiatry* 16 (2): 181–202.

Pogliano, Claudio. 2004. "Sciences at War and the Cybernetic Dream." *Nuncius* 19:171–204.

Pollock, Horatio M. 1939. "A Statistical Study of 1,140 Dementia Praecox Patients Treated with Metrazol." *Psychiatric Quarterly* 13:558–568.

Pols, Hans. 2013. "Beyond the Clinical Frontiers: The American Mental Hygiene Movement, 1910–1945." In *International Relations in Psychiatry: Britain, Germany, and the United States to World War II*, ed. Volker Roelke, Paul J. Weindling, and Louise Westwood, 111–133. Rochester, NY: University of Rochester Press.

Pols, Hans. 2011. "The Tunisian Campaign, War Neuroses, and the Reorientation of American Psychiatry during World War II." *Harvard Review of Psychiatry* 19:313–320.

Pols, Hans. 2007. "War Neurosis, Adjustment Problems in Veterans, and an Ill Nation: The Disciplinary Project of American Psychiatry during and after World War II." *Osiris* Second Series 22:72–92.

Pressman, Jack D. 1998. "Human Understanding: Psychosomatic Medicine and the Mission of the Rockefeller Foundation." In *Greater than the Parts: Holism in Biomedicine, 1920–1950*, ed. Christopher Lawrence and George Weisz, 189–208. New York: Oxford University Press.

Pressman, Jack D. 1998. *Last Resort: Psychosurgery and the Limits of Medicine.* New York: Cambridge University Press.

Pressman, Jack D. 1988. "Sufficient Promise: John F. Fulton and the Origins of Psychosurgery." *Bulletin of the History of Medicine* 62:1–22.

Pribram, Karl. 1966. "Brilliant, Lively, Productive, and Wrong." *PsycCRITIQUES* 11 (12): 580–581.

Psillios, Stathis. 2011. "An Explorer upon Untrodden Ground: Peirce on Abduction." In *The Handbook of the History of Logic, Volume 10: Inductive Logic*, ed. Dov M. Gabbay, Stephan Hartmann, and John Woods, 117–151. North Holland.

Rall, Wilfrid. 1990. "Some Historical Notes." In *Computational Neuroscience*, ed. E. L. Schwartz, 3–8. Cambridge, MA: MIT Press.

Rashevsky, Nicolas. 1938. *Mathematical Biophysics: Physicomathematical Foundations of Biology*. Chicago: University of Chicago Press.

Rashevsky, Nicolas. 1933. "Outline of a Physico-Mathematical Theory of Excitation and Inhibition." *Protoplasma* 20:42–56.

Rashevsky, Nicolas. 1931. "Possible Brain Mechanisms and Their Physical Models." *Journal of General Psychology* 5:368–406.

Rashevsky, Nicolas. 1931. "On the Theory of Nerve Conduction." *Journal of General Physiology* 14:517–528.

Rasmussen, Nicolas. 2002. "Of 'Small Men,' Big Science and Bigger Business: The Second World War and Biomedical Research in the United States." *Minerva: A Review of Science Learning and Policy* 40:115–146.

Reingold, Nathan. 1995. "Choosing the Future: The US Research Community: 1944–46." *Historical Studies in the Physical and Biological Sciences* 25:301–328.

Reingold, Nathan. 1994. "Science and Government in the United States since 1945." *History of Science* 32:361–386.

Reingold, Nathan. 1979. *The Sciences in the American Context: New Perspectives*. Washington, DC: Smithsonian Institute Press.

Reisch, George A. 2005. *How the Cold War Transformed Philosophy of Science: To the Icy Slopes of Logic*. Cambridge: Cambridge University Press.

"Research Groups Listed in R.L.E.—Q.P.R. at 5-Year Intervals, Research Laboratory of Electronics," *R.L.E.: 1946+20*, Research Laboratory of Electronics, Massachusetts Institute of Technology, Cambridge, MA, May 1966, 24–25.

Resnick, David. 1991. "How-Possibly Explanations in Biology." *Acta Biotheoretica* 39 (2): 141–149.

Rheinberger, Hans-Jörg. 1997. *Toward a History of Epistemic Things: Synthesizing Proteins in the Test Tube*. Stanford, CA: Stanford University Press.

Rheinberger, Hans-Jörg. 1992. "Experiment, Difference, and Writing. Part 1. Tracing Protein Synthesis. Part 2. The Laboratory Production of Transfer RNA." *Studies in History and Philosophy of Science* 23:305–331, 389–422.

Richards, Joan L. 2006. "Introduction: Fragmented Lives." *Isis* 97:302–305.

Richardson, Alan. 2013. "Philosophy of Science in America." In *The Oxford Handbook of American Philosophy*, ed. Cheryl Misak, 1–53. Oxford Handbooks Online, Oxford University Press, DOI: 10.1093/oxfordhb/9780199219315.003.0017.

Richardson, Alan. March 2008. "Scientific Philosophy as a Topic for History of Science." *Isis* 99 (1): 88–96.

Richardson, Alan W. 2003. "Logical Empiricism, American Pragmatism, and the Fate of Scientific Philosophy in North America." In *Logical Empiricism in North America, Minnesota Studies in the Philosophy of Science*, vol. XVIII, ed. Gary L. Hardcastle and Alan W. Richardson, 1–24. Minneapolis: University of Minnesota Press.

Richardson, Alan. 2003. "The Geometry of Knowledge: Lewis, Becker, Carnap, and the Formalization of Philosophy in the 1920s." *Studies in History and Philosophy of Science* 34:165–182.

Richardson, Alan W., and Gary L. Hardcastle. 2003. "Introduction: Logical Empiricism in North America." In *Logical Empiricism in North America, Minnesota Studies in the Philosophy of Science*, vol. XVIII, ed. Gary L. Hardcastle and Alan W. Richardson, vii–xxix. Minneapolis: University of Minnesota Press.

Robin, Ron. 2011. *Making the Cold War Enemy: Culture and Politics in the Military-Industrial-Academic Complex*. Princeton, NJ: Princeton University Press.

Rosenberg, Charles. 1979. "Toward an Ecology of Knowledge: On Discipline, Context, and History." In *The Organization of Knowledge in Modern America, 1860–1920*, ed. Alexandra Oleson and John Voss, 440–455. Baltimore, MD: Johns Hopkins University Press.

Rosenblueth, Arturo, and Norbert Wiener. 1945. "The Role of Models in Science." *Philosophy of Science* 12:316–321.

Rosenblueth, Arturo, Norbert Wiener, and Julian Bigelow. 1943. "Behavior, Purpose, and Teleology." *Philosophy of Science* 10:18–24.

Rothstein, William G. 1987. *American Medical Schools and the Practice of Medicine: A History*. New York: Oxford University Press.

Russell, Bertrand. 1959. "Logical Atomism." In *Logical Positivism*, ed. Alfred J. Ayer, 31–50. Glencoe, IL: Free Press.

Russell, Bertrand. 1920. *Introduction to Mathematical Philosophy*. London: George Allen and Unwin.

Sachs, E. 1952. *The History and Development of Neurological Surgery*. New York: Paul Hoeber.

Sadowsky, Jonathan. 2005. "Beyond the Metaphor of the Pendulum: Electroconvulsive Therapy, Psychoanalysis, and the Styles of American Psychiatry." *Journal of the History of Medicine and Allied Sciences* 61 (1): 1–25.

Sapolsky, Harvey M. 1990. *Science and the Navy: The History of the Office of Naval Research*. Princeton, NJ: Princeton University Press.

Schmidt-Brücken, Katharina. 2012. *Hirnzirkel: Kreisende Prozesse in Computer und Gehirn: Zur neurokybernetischen Vorgeschichte der Informatik*. Bielefeld: Transcript.

Schneider, William H. 2002. "The Men Who Followed Flexner: Richard Pearce, Alan Gregg, and the Rockefeller Foundation Medical Divisions, 191–1951." In *Rockefeller Philanthropy and Modern Biomedicine: International Initiatives from World War I to the Cold War*, ed. William H. Schneider, 7–60. Bloomington: Indiana University Press.

Schrecker, Paul. 1947. "Leibniz and the Art of Inventing Algorisms." *Journal of the History of Ideas* 8:107–116.

Scull, Andrew. April 10, 2010. "A Psychiatric Revolution." *Lancet* 275 (9722): 1246–1247.

Scull, Andrew. 2005. *Madhouse: A Tragic Tale of Melagomania and Modern Medicine*. New Haven: Yale University Press.

Sealander, Judith. 1997. *Private Wealth and Public Life: Foundation Philanthropy and the Reshaping of American Social Policy from the Progressive Era to the New Deal*. Baltimore/London: Johns Hopkins University Press.

Sejnowski, T. J., C. Koch, and P. S. Churchland. 1988. "Computational Neuroscience." *Science* 241:1299–1306.

Shail, Andrew, and Laura Salisbury, eds. 2010. *Neurology and Modernity: A Cultural History of Nervous Systems, 1800–1950*. New York:Palgrave Macmillan.

Shepherd, Gordon M. 2010. *Creating Modern Neuroscience: The Revolutionary 1950s*. New York: Oxford University Press.

Sherrington, Charles S. 1947. *Integrative Action of the Nervous System, New Edition*. Edinburgh: Cambridge University Press.

Sherrington, Charles S. 1940. *Man on his Nature. The Gifford Lectures*, 1937–1938. Edinburgh: Cambridge University Press.

Sherrington, Charles S. 1906. *The Integrative Action of the Nervous System*. New York: Charles Scribner's Sons.

Shmailov, Maya M. 2013. "Nicolas Rashevsky's Pencil-and-Paper Biology." In *Outsider Scientists: Routes to Innovation in Biology*, ed. Oren Harman and Michael R. Dietrich, 161–180. Chicago: University of Chicago Press.

Shorter, Edward, and David Healy. 2007. *Shock Therapy: A History of Electroconvulsive Treatment in Mental Illness*. Toronto: University of Toronto Press.

Shortland, Michael, and Richard Yeo, eds. 1996. *Telling Lives in Science: Essays on Scientific Biography*. Cambridge: Cambridge University Press.

Simon, Jonathan. 1999. "Naming and Toxicity: A History of Strychnine." *Studies in History and Philosophy of Biological and Biomedical Sciences* 30 (4): 505–525.

Sloan, Philip R. September 2014. "Molecularizing Chicago—1945–1965: The Rise, Fall, and Rebirth of the University of Chicago's Biophysics Program." *Historical Studies in the Natural Sciences* 44 (4): 364–412.

Smalheiser, Neil R. 2000. "Walter Pitts." *Perspectives in Biology and Medicine* 43 (2): 217–226.

Smith, Laurence D. 1990. "Metaphors of Knowledge and Behavior in the Behaviorist Tradition." In *Metaphors in the History of Psychology*, ed. D. E. Leary, 237–266. Cambridge: Cambridge University Press.

Smith, Laurence D. 1986. *Behaviorism and Logical Positivism: A Reassessment of the Alliance*. Stanford, CA: Stanford University Press.

Smith, Roger. 2009. "Review of Kenton Kroker, *The Sleep of Others and the Transformation of Sleep Research*." *History of the Human Sciences* 22 (5): 108–113.

Smith, Roger. 2005. "The History of Psychological Categories." *Studies in History and Philosophy of Science Part C, Studies in History and Philosophy of Biological and Biomedical Sciences* 36:55–94.

Smith, Roger. 2001. "Physiology and Psychology, or Brain and Mind, in the Age of C. S. Sherrington." In *Psychology in Britain: Historical Essays and Personal Reflections*, ed. G. C. Bunn, A. D. Lovie, and G. D. Richards, 223–242. Leicester: British Psychological Society.

Smith, Roger. 2001. "Representations of Mind: C. S. Sherrington and Scientific Opinion, ca. 1930–1950." *Science in Context* 14 (4): 511–539.

Smith, Roger. 2000. "The Embodiment of Value: C. S. Sherrington and the Cultivation of Science." *British Journal for the History of Science* 33(3), no. 118: 283–311.

Smith, Roger. 1997. *The Human Sciences*. New York: W.W. Norton.

Smith, Sidonie, and Julia Watson. 2010. *Reading Autobiography: A Guide for Interpreting Life Narratives*. Minneapolis, MN: University of Minnesota Press.

Söderqvist, Thomas. 2007. "Introduction: A New Look at the Genre of Scientific Biography." In *The History and Poetics of Scientific Biography*, ed. Thomas Söderqvist, 1–15. Aldershot, UK: Ashgate.

Sokal, Michael M. 1984. "James McKeen Cattell and American Psychology in the 1920s." In *Explorations in the History of Psychology in the United States*, ed. Josef Brožek,

273–323. Lewisburg, PA: Bucknell University Press; London: Associated University Presses.

Solomon, C., S. Cobb, and W. Penfield, eds. 1958. *The Brain and Human Behavior: Proceedings of the Meeting of the Association for Research in Nervous and Mental Disease, 7 and 8 December, 1956.* Baltimore, MD: Williams and Wilkins.

Solovey, Mark. 2013. *Shaky Foundations: The Politics-Patronage-Social Science Nexus in Cold War America.* New Brunswick, NJ: Rutgers University Press.

Solovey, Mark. 2001. "Science and the State during the Cold War: Blurred Boundaries and a Contested Legacy." *Social Studies of Science* 31 (2): 165–170.

Squire, Larry R., ed. 1996. *The History of Neuroscience in Autobiography,* vol. I. Washington, DC: Society for Neuroscience.

Stanley, Matthew. 2015. *Huxley's Church and Maxwell's Demon: From Theistic Science to Naturalistic Science.* Chicago: University of Chicago Press.

Star, Susan Leigh. 1989. *Regions of the Mind: Brain Research and the Quest for Scientific Certainty.* Stanford, CA: Stanford University Press.

Star, Susan Leigh. 1986. "Triangulating Basic and Clinical Research: British Localizationists, 1870–1906." *History of Science* 24:29–48.

Starks, Sarah Linsley, and Joel T. Braslow. 2005. "The Making of Contemporary American Psychiatry, Part 1: Patients, Treatments, and Therapeutic Rationales before and after World War II." *History of Psychology* 8 (2): 176–193.

Stratton, Julius A. 1966. "RLE: The Beginning of an Idea." In *Research Laboratory of Electronics, R.L.E.: 1946+20,* Research Laboratory of Electronics, Massachusetts Institute of Technology (Cambridge, MA, May 1966), 1–6.

Sulloway, Frank. 1979. *Freud: Biologist of the Mind.* New York: Basic Books.

Taylor, Eugene. 1999. *Shadow Culture: Psychology and Spirituality in America.* Washington, DC: Counterpoint.

Terrall, Mary. 2014. *Catching Nature in the Act: Réaumur and the Practice of Natural History in the Eighteenth Century.* Chicago/London: University of Chicago Press.

Terrall, Mary. 2006. "Biography as Cultural History of Science." *Isis* 97:306–313.

Terrall, Mary. 2002. *The Man Who Flattened the Earth: Maupertuis and the Sciences in the Enlightenment.* Chicago: University of Chicago Press.

Thelin, John R. 2011. *A History of American Higher Education.* 2nd ed. Baltimore, MD: Johns Hopkins University Press.

Thorne, Frederick C. 1976. "Reflections on the Golden Age of Columbia Psychology [1920–1940]." *Journal of the History of the Behavioral Sciences* 12:159–165.

Todes, Daniel. 2002. *Pavlov's Physiology Factory: Experiment, Interpretation, Laboratory Enterprise*. Baltimore, MD: Johns Hopkins University Press.

Tomes, Nancy. 2000. "Introduction." *Bulletin of the History of Medicine* 74 (3), 773–777.

Turing, Alan M. 1936–1937. "On Computable Numbers, with an Application to the *Entscheidungsproblem*." *Proceedings of the London Mathematical Society, Series 2*, 42: 230–265.

Valenstein, Elliot. 1986. *Great and Desperate Cures: The Rise and Decline of Psychosurgery and Other Radical Treatments for Mental Illness*. New York: Basic Books.

Volterra, Vito. 1931. *Leçons sur la Théorie Mathématique de la Lutte pour la Vie*. Paris: Gauthier-Villars et cie.

Von Bonin, Gerhardt, Hugh W. Garol, and Warren S. McCulloch. 1942. "Functional Organization of the Occipital Lobe." In *Visual Mechanisms (Biological Symposia Volume VII)*, ed. Heinrich Klüver, 65–192. Lancaster, PA: Jacques Cattell Press.

Von Domarus, Eilhard. 1944. "The Specific Laws of Logic in Schizophrenia." In *Language and Thought in Schizophrenia: Collected Papers Presented at the Meeting of the American Psychiatric Association, May 12, 1939*, ed. J. S. Kasanin, 104–114. Berkeley, CA: University of California Press.

Von Domarus, Eilhard. 1967/1930. "The Logical Structure of Mind: An Inquiry into the Philosophical Foundation of Psychology and Psychiatry." In *Communication: Theory and Research. Proceedings of the First Symposium*, ed. L. Thayer, 351–411. Springfield, IL: Charles C. Thomas.

Von Domarus, Eilhard. 1929. *Das Denken und seine krankhaften Störungen*. Leipzig: Kabitzsch.

Von Domarus, Eilhard. 1928. "Über das Denken der Manischen und Depressiven." *Zeitschrift für die Gesamte Neurologie und Psychiatrie* 112:632–635.

Von Domarus, Eilhard. 1927. "Zur Theorie des schizophrenen Denkens." *Zeitschrift für die Gesamte Neurologie und Psychiatrie* 108:703–714.

Von Domarus, Eilhard. 1923. "Prälogisches Denken in der Schizophrenie." *Zeitschrift für die Gesamte Neurologie und Psychiatrie* 87 (1): 84–93.

Von Eckardt, B. 1993. *What Is Cognitive Science?* Cambridge, MA: MIT Press/Bradford Books.

Von Foerster, Heinz. 1989. "Circular Causality: The Beginnings of an Epistemology of Responsibility." In *The Collected Works of Warren S. McCulloch*, vol. 3, ed. Rook McCulloch, 808–829. Salinas, CA: Intersystems.

Von Foerster, Heinz, and George W. Zopf, Jr., eds. 1962. *Principles of Self-Organiza-tion: Transactions of the University of Illinois Symposium on Self-Organization, Robert Allerton Park, 8 and 9 June, 1961.* New York: Pergamon Press.

Von Neumann, John. 2000. *The Computer and the Brain (Yale University, Mrs. Hepsa Ely Silliman Memorial Lectures, 1958).* New Haven/London: Yale University Press.

Von Neumann, John. 1951. "The General and Logical Theory of Automata." In *Cerebral Mechanisms in Behavior: The Hixon Symposium,* ed. Lloyd A. Jeffress, 1–31. New York: John Wiley.

Von Neumann, John. 1945. "First Draft Report on the EDVAC." Report prepared for the U.S. Army Ordnance Department under contract W-670-ORD-4926. In N. Stern, *From ENIAC to UNIVAC,* 177–246. Bedford, MA: Digital Press, 1981.

Walter, W. Grey. 1966. "Review of Warren S. McCulloch, *Embodiments of Mind.*" *Electroencephalography and Clinical Neurophysiology* 21:311–312.

Wanke, Paul. 1999. "American Military Psychiatry and Its Role among Ground Forces in World War II." *Journal of Military History* 63 (1): 127–146.

Ward, Arthur A., and Horace W. Magoun. 1948. "Production of an Alternating Tremor at Rest in Monkeys." *Journal of Neurophysiology* 11:317–330.

Ward, Arthur A., Warren S. McCulloch, and N. Kopeloff. 1948. "Temporal and Spa-tial Distribution of Changes during Spontaneous Seizures in Monkey Brain." *Journal of Neurophysiology* 11: 377–386.

Weidman, Nadine M. 1999. *Constructing Scientific Psychology: Karl Lashley's Mind-Brain Debates.* Cambridge University Press.

Weisz, George. 2006. *Divide and Conquer: A Comparative History of Medical Specializa-tion.* Oxford University Press.

Wheatley, M. D., and Warren S. McCulloch. 1947. "Sundry Changes in Physiology of Cerebral Cortex Following Rapid Injection of Sodium Cyanide." *Federal Proceed-ings of the American Societies for Experimental Biology* 6, no. 1 225.

White, Mark Andrew. 2000. "Hambidge, Jay." *American National Biography Online.* http://www.anb.org.myaccess.library.utoronto.ca/articles/17/1700364.html

Wiener, Norbert. 1950. *The Human Use of Human Beings.* Cambridge, MA: MIT Press.

Wiener, Norbert. 1948. *Cybernetics: Or Control and Communication in the Animal and the Machine.* Cambridge, MA: MIT Press.

Wiener, Norbert, and Arturo Rosenblueth. 1946. "The Mathematical Formulation of the Problem of Conduction of Impulses in a Network of Connected Excitable Elements, Specifically in Cardiac Muscle." *Archivos del Instituto de Cardiologia de Mexico*::205–265.

Wiesner, Jerome B. May 1966."The Communication Sciences—Those Early Days." In *Research Laboratory of Electronics, R.L.E.: 1946+20*, Research Laboratory of Electronics, Massachusetts Institute of Technology, Cambridge, MA, 12–16.

Wilson, Daniel J. 1990. *Science, Community, and the Transformation of American Philosophy, 1860–1930*. Chicago/London: University of Chicago Press.

Wilson, Elizabeth A. 2010. *Affect and Artificial Intelligence*. Seattle/London: University of Washington Press.

Winter, Alison. 2012. *Memory: Fragments of a Modern History*. Chicago: University of Chicago Press.

Worden, Frederic, Judith P. Swazey, and George Adelman, eds. 1975. *The Neurosciences: Paths of Discovery*. Cambridge, MA: MIT Press.

Wortis, S. Bernard, and Warren S. McCulloch. 1932. "Head Injuries: An Experimental Study." *Archives of Surgery* 25:529–543.

Young, Robert M. 1970. *Mind, Brain, and Adaptation in the Nineteenth Century: Cerebral Localization and Its Context from Gall to Ferrier*. Oxford: Clarendon Press.

Zuckerman, S. 1950. *The Physical Basis of Mind*. Ed. Peter Laslett. Basil Blackwell.

Zunz, Olivier. 2012. *Philanthropy in America: A History*. Princeton, NJ: Princeton University Press.

# Index